ECHOES OF EARTH

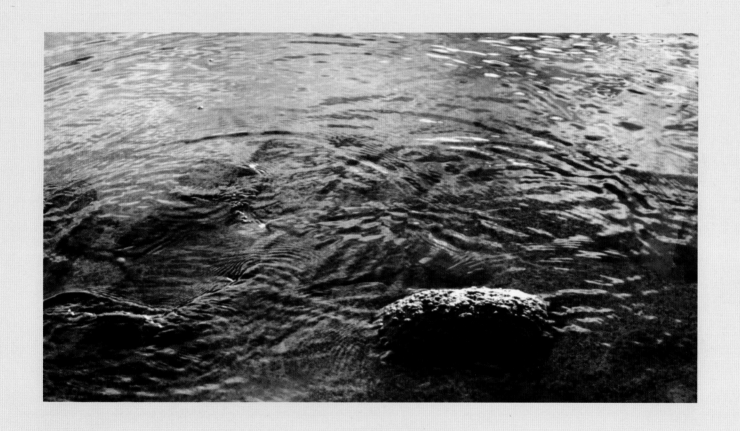

ECHOES OF EARTH

L. SUE BAUGH

Finding Ourselves

in the Origins of the Planet

Designer: Jericho Hernandez,
  Thikeworks, www.jerichohernandez.com
Research and Permissions: Fran Brown
Copyeditors: Judith Gallagher, Sue Roupp
Proofreader: Dee Stiffler
Illustrators: Sioban Lombardi, L. Sue Baugh

Technical Assistance:
Drum scanning: Mat Lombardi and Dustin
Johnson of Pixel Mint, Chicago, IL
Photo printing: Chris Linster and James
Aitken, Quartet Copies, Evanston, IL;
MotoPhoto, Evanston, IL; Kodak Company,
Switzerland
Photo research: Photo Researchers,
New York, NY

Library of Congress Control Number:
2011935197

*Front cover and full title page: Akilia Island,
Greenland*
*Back cover and half title page: Hamelin Pool,
Shark Bay, Western Australia*

First Printing.

Publisher Cataloging-in-Publication Data

Baugh, L. Sue
Echoes of Earth: Finding Ourselves in the
Origins of the Planet/L. Sue Baugh, 212 p.:
Includes index. Summary: Documents some
of the world's oldest rock and minerals
sites and shows how we carry the long
history of Earth within our bodies.

ISBN-978-0-9838576-3-1 (hardcover)

1. Spirituality 2. Nature 3. Travel 4. Earth
Science I Title

Published by

Wild Stone Arts™
www.wildstonearts.com

Printed and bound in China by Hung Hing
          www.hunghingprinting.com

TO THE READER

from Lynn Martinelli

The book you hold in your hands is the image and voice of a journey that took 10 years and nearly 54,000 miles to accomplish—and still continues. For me, this journey seemed to reset my clock to zero, to a true and essential moment of being. I revisit that moment often, never failing to return recharged. The photos of the rocks keep me ticking in rhythm with Nature.

The unique treasure of our travels, however, is what we affectionately call "the story of the stones." Sue and I arrived at each site and embraced its solitude, timelessness, and silence. We rarely spoke; we listened. Working from an artistic perspective, we let go of the need to record scale or dimension and gained instead a different kind of knowledge.

Long after our final trip, we sat amidst our accumulated photos and notebooks strewn about like pieces of a huge jigsaw puzzle and asked: What makes these remote places so true that they are able to touch our deepest core; so powerful that they catalyzed our own spiritual and artistic growth; and yet so welcoming that we felt utterly enfolded in the most bountiful and nurturing of homes?

In putting this book together, Sue has found an inspired way to recount the complex and simple threads of the story of the stones. She not only created innovative book layouts to convey our changing perspective but also merged science and art, nature and human, personal and universal. These connections may surprise some of you while others of you may already sense them.

We hope this book will be your first step toward discovering these places in the world and in yourselves.

Yorro, Yorro—everything standing up alive
—Australian Aboriginal saying

TABLE OF

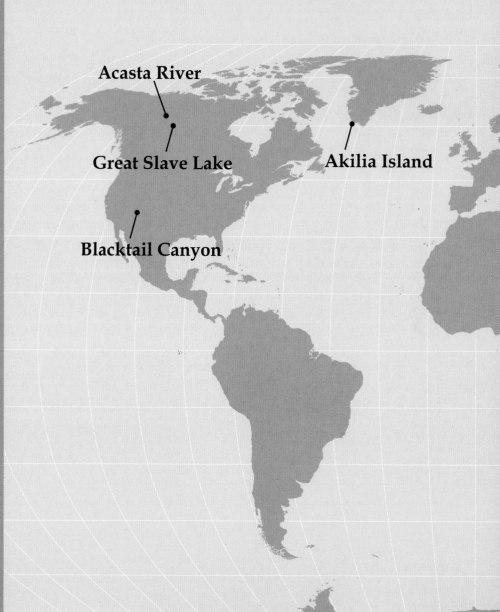

Acasta River

Akilia Island

Great Slave Lake

Blacktail Canyon

CONTENTS

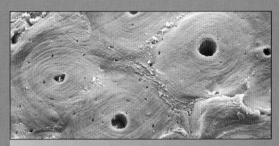

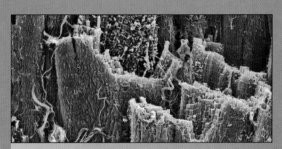

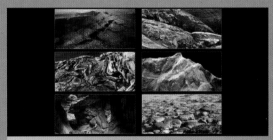

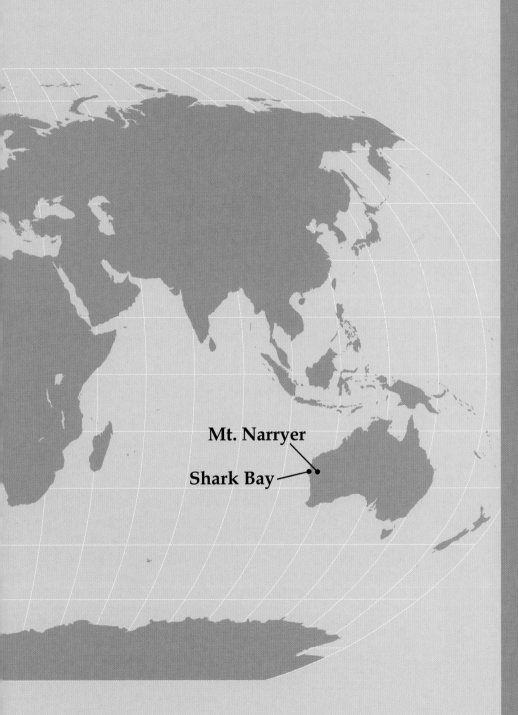

Mt. Narryer

Shark Bay

FOREWORD

This book began as a creative project to document the world's oldest rock and mineral sites. As artists, we sought to experience a landscape that echoed what ancient Earth might have been like long before humanity appeared.

Our search led us into remote regions of Western Australia, Canada, Greenland, the United States... and eventually into territory not found on any map.

Far from finding a landscape empty of humankind, we discovered that our very beginnings lie hidden in the story of the oldest stones. The power and beauty of these ancient sites radically changed our perception of the world and humanity's place in it.

We invite you to explore the mystery

of what we encountered—

that we carry the long history

of the planet within us

*. . . we
are all
Echoes of
Earth.*

*The journey starts in the
Australian Outback.*

(Over)

MT. NARRYER

26° 35′ 13″ S
115° 55′ 32″ E

Indian Ocean

• Perth

MT. NARRYER

WESTERN AUSTRALIA

The story begins in the stones . . .

MT. NARRYER

WESTERN AUSTRALIA

Mt. Narryer turns blood red at sunrise, its rippled earth looking like waves frozen in time. The mountain was thrust up by the immense, grinding collision of landmasses that formed Australia. Enfolded in its rocks are tiny crystals of zircon whose layers carry the story of Earth's Deep Time, reaching back over four billion years to the beginning of the planet itself.

But more than crystal memory keepers compels us here. We have crossed the Pacific Ocean, flown to the west coast of Australia, and driven 300 miles into the Outback, drawn by the vast solitude surrounding this ancient mountain.

Out here there is no cell phone service, no GPS, no Internet. We bring only our notebooks, cameras, and a willingness to listen deeply. In many ways, we are both ready and totally unprepared for this encounter.

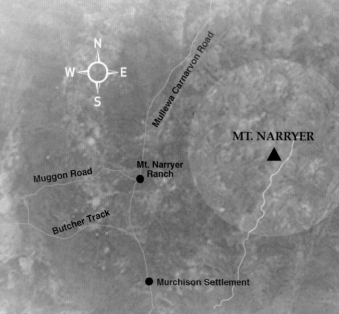

© Reg Morrison

We come to the site as visitors,

seeking a knowledge lost from our lives.

We will leave knowing irrevocably

we are children of this place.

THE STORY BEGINS
IN THE STONES

*O*nly three days earlier we were
in Perth, sitting in the office of Dr. David Nelson, an expert on Mt. Narryer at Curtin University. We'd arrived in July, the heart of a winter far milder than our winters in Chicago. Perth spreads along the banks of the great Swan River and is graced by warm winds off the Indian Ocean. We had not expected to find a landscape so filled with light.

"How can I help you?" Dr. Nelson leans back in his chair and regards us with curiosity. His small office is crowded with papers, books, and rock specimens. We decide to tell the simple truth. "We'd like your help to visit Mt. Narryer as part of our project to

document the world's oldest rock and mineral sites. But we're not doing a scientific piece. We want to approach these sites as artists—no preconceptions, just visit the land and spend time listening. We have no idea what we'll find there."

As we talk, Dr. Nelson's eyes narrow and our hearts sink. What must his academic mind be thinking? He leans forward slowly, his voice quiet. "You know, I try to get my students to do the same thing. I tell them, don't just study the site; take time to commune with it." Halfway around the world, in the most unlikely place, we have found a kindred spirit.

Tiny zircons, smaller than grains of sand, shimmer in polarized light. As these crystals grow, their layers trap radioactive uranium, which decays at a steady rate into lead. The higher the ratio of lead to uranium, the older the zircon. Zircons found at Mt. Narryer and Jack Hills were dated at 3.6 to 4.4 billion years old.

When we ask about traveling in the Outback, he warns us, "Cell phones don't work out there, and you won't have Internet service. The car rental will give you a shortwave radio to call for help, but the nearest doctor is about 200 miles away by plane."

Before we leave, Dr. Nelson shows us the research laboratory where he and his colleagues first discovered the immense age of the tiny zircons from Mt. Narryer and Jack Hills to the north. "When the dating sequence finished, it showed we had a zircon that was 4.2 billion years old—almost the age of Earth! It made my hair stand on end."

The McTaggarts' Mountain

Dr. Nelson gives us the phone number of Carol and Sandy McTaggart, who own the land around Mt. Narryer. "They can guide you right to the mountain. Carol is known as the 'Keeper of the Mount'—she's trying to protect Mt. Narryer as a geological site."

When we call Carol, asking for permission to visit Mt. Narryer, she warmly invites us to come as if we were old friends. That afternoon we rent our camper car—equipped with two beds, a small kitchen, and the essential shortwave radio—and head straight north for Mt. Narryer Ranch.

For a day and a half, we drive over Western Australia's graded red dirt roads—there are no paved roads away from the coast. We are struck by the vivid blue of the sky, stark white trunks of sweetgum trees, and harlequin colors of parrots, budgies, and magpies flashing in the branches. By the time we turn off Mullewa Carnarvon Road to Mt. Narryer Ranch, the late afternoon sun is casting long, thin shadows over the red earth. Carol meets us as we drive up and then ushers us into the ranch house where Sandy is waiting. "Sorry, we can't take you to the mount," he says, "but we're starting a big roundup of our sheep tomorrow."

Sandy sketches a map for us, marking all the turnoffs, gates, fences, and rough bits between the ranch house and Mt. Narryer. Carol adds, "There's enough daylight left that you can make Jailor's Camp by tonight—that's the halfway mark." Sandy cautions us, "Remember, if you get lost, we can't even begin to look for you for four days."

Carol lends us a few blankets for the chilly nights, and Sandy calls, "Don't get lost!" from the doorway. We start out on our adventure, driving away from the ranch on a dirt road wide as a four-lane highway. In our excitement to reach the mountain, we don't realize that we haven't brought enough food and water with us. Our naïve, city-trained minds must have assumed we would run across a convenience store somewhere in the Outback.

THE ROADS TO MT. NARRYER

As twilight deepens, we pull into Jailor's Camp and spend our first cold night in the Outback. When we start out early the next morning, we quickly find ourselves in trouble. Not only has our wide dirt road narrowed to a single lane, it's no longer taking us toward the mountain.

For two hours we drive back and forth from Jailor's Camp, trying to find the right way. We finally stop, get out of the car, and study the rough map Sandy sketched for us. It clearly shows only one route to Mt. Narryer, so why does it keep taking us south instead of east where the mountain lies?

We try to spot a road in the red, arid land around us, dotted with shrubs and shaded by eucalyptus and acacia trees. Then something inside us shifts. We stop looking for a "road" and simply gaze at the land . . . *The way is here somewhere; what are we missing?*

And then we see it: two faint tire tracks gleaming in the bent grass. The morning sun has reached just the right angle to reflect off the golden stalks. There, curving to the right like a faded ribbon, lies our road to Mt. Narryer.

Jailor's Camp was once a holding area for prisoners being taken to the nearest town. It lies midway between the McTaggarts' ranch and Mt. Narryer.

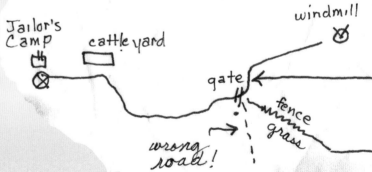

▲ Our first glimpse of Mt. Narryer in the morning light.

Black Hill →

Cairn

mesA at Mt. Narryer

← 6 km →

rough bits

this is what "rough bits" → meant!

gate

fence

The closer we travel to the mountain, the more the land seems to move back in time. Soon the ranch buildings and fences lie far behind us. At one point, even the road disappears into the "rough bits"—rocky creek beds lined with brush and scrub trees. Taking a page from Hansel and Gretel, we tie black plastic strips to the bushes so we can find our way back to the main road again.

The base of Mt. Narryer looms

before us as we finally reach the mountain and park our camper near a small red mesa. The ground is littered with rocks eroded from the mountain's steep sides, forcing us to step carefully. The stones give off a dry acrid smell in the cool air.

We're not sure how or where to begin our work. The only living things we see around us are magpies, white moths, and a few dragonflies hovering over the dry brush. A magpie's call rings through the air like notes from a glass flute.

A Language Older Than Words

We're startled to hear the sound of traffic on a distant highway intruding into this complete solitude. Then reality hits: there are no highways for miles in any direction.

That sound we hear is the wind moving through groves of acacia trees. At that moment, the voice of this place asserts itself, speaking in a language older than words. Like others before us, we are learning how to listen.

> ## " We're in the middle of everywhere. "

The next morning we decide to

climb Mt. Narryer and photograph its varied faces. The mountain stands only 1100 feet high but has no trails up its side. We struggle to pick our way over tumbled boulders and jagged, broken rocks.

Carol had told us, "Be careful—even geologists get lost in the washes and ravines." We move slowly, rationing our food and water. Each time we hit a dead end, we stop and let our eyes shift and a path opens up, as if the mountain were guiding us. Far below, our car becomes a small white spot on the landscape.

At noon, we finally reach the top and see the mountain curve away from the summit like the body of a small dragon. Everywhere we look, the Outback stretches as flat and wide as an ocean. Here the Aboriginal six directions make perfect sense: north, south, east, west, *above* and *below*.

We're in the middle of everywhere.

This sandstone formation stands

like a sentinel guarding the summit. A few granite boulders lie scattered about, survivors of even older mountain ranges that arose and eroded away eons ago.

Here we feel Earth's Deep Time rise up through the ground like heat, infusing this place with a profound sense of clarity and peace. We photograph roots pushing through solid rock and the intricate patterns carved in stone by wind and water over millions of years. The mountain's more intimate faces are teaching us a new meaning of the word "slow."

When we return to our camp

something within us has changed. We talk less, move more slowly, listen with more patience. By the second day, our sketchbooks and journals have few entries as photography becomes our primary language. When we rest atop the mesas, our human forms seem to echo the shape and color of stone.

The more we listen, the more we see, as if we had thrown a rock into the silent core of the mountain and the soundless ripples slowly make visible the life around us and within us.

We watch as spiders build webs in ravines; each day, downdrafts blow insects into their traps. Ants and rodents erect small dikes around their tunnels to keep winter rains from flooding them. Like the tiny four-billion-year-old zircons, every plant and animal here is a master of survival. In this land of drought and flood, there are few second chances.

Yet even though we're short of water and miles from any quick rescue, there's an odd feeling of protection here. This mountain shelters life as well as tests it.

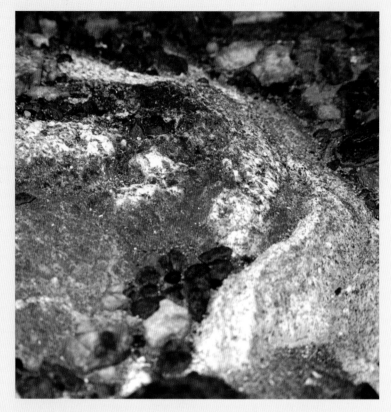

Brilliant abstract patterns appear in the ground around Mt. Narryer, created as seasonal rains leach chemicals from the stones.

Shrubs, silvery in the low winter light, send out roots in thick, underground mats that keep the fragile soil from eroding.

Goat tracks mark the earth like cryptic runes. The goats escaped from their European owners and now thrive in this semiarid land.

Quartz clusters lie exposed in the dry air as softer stone around them erodes. Drab by day, these stones in moonlight glow with a white fire.

We found bleached bones of small rodents scattered around the summit, victims of raptors and poisonous snakes.

This succulent plant looks soft as velvet, but tiny, needle-sharp spines cover its leaves for protection and to conserve water.

Rodents use stones to anchor their small dikes, built of dry grass and mud, which prevent runoff from Mt. Narryer from flooding their homes.

On our last night at Mt. Narryer,

we build a fire in the chill air and share what's left of our food. The mountain flares like a fiery ember as the sun sinks lower.

Beneath the apparent harshness of this land, an ancient web lies unbroken and welcomes two of its youngest children as its own. We are only beginning to plumb the depths of it.

Our fire dwindles, and darkness slowly reveals the Milky Way arching above like a clouded path; the dome of sky fills with stars. All around us, quartz stones glow in the moonlight, mirroring the starry world overhead.

Held between this primal earth and sky, we have never felt so much at home.

> *"We don't own the land; we belong to it."*
> —*Australian Aboriginal saying*

KEEPERS OF THE MOUNTAIN

"You want to see Mt. Narryer? You have to talk to the Keeper of the Mount."

—Sandy McTaggart

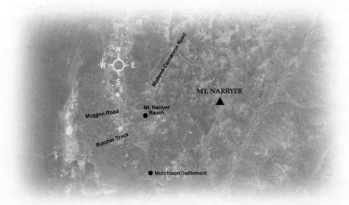

"*What did you find out there?*"

Carol is eager to hear about our stay at her beloved mountain. We returned to the McTaggarts' house at dusk, having followed our plastic strips back through the bush until we found the main road again. Now Carol and the two of us warm ourselves near the hearth where a single log burns in the fire. Carol says, "A Japanese visitor once told me Mt. Narryer has *shibumi*, a kind of condensed power."

"That's what we felt, too," we tell her, "and it's probably going to take us years to understand it." As the fire burns lower, we ask how they came to own Mt. Narryer. Carol laughs. "I knew it was something special the first time I climbed it. I asked Sandy to buy the mountain for me—and he did!

"My first goal in life was to be an anthropologist, but in the 1960s, when I got my degree, women weren't allowed in the field with the men. They told me I could teach or catalog other people's work. So instead, I studied city planning, traveled to Australia where I met Sandy, and here I am!"

After the zircons were discovered at Mt. Narryer, Carol tells us, geology teams from all over the world started coming to the ranch. "In fact, we took a NASA team out to Mt. Narryer a week before you came." Now, in a wonderful twist of fate, Carol not only guides expeditions to the mountain but also owns the land where all the fieldwork is done.

The Spirit of Place

For the McTaggarts, Mt. Narryer is more than a geologic mecca. It serves as a refuge from the hard life of ranching in the Outback, allowing them to touch something enduring and sustaining. As Sandy has said, "You can commune there."

◄ Carol and Sandy McTaggart on top of Mt. Narryer. They have worked tirelessly to protect their mountain.

Mt. Narryer rises like the back of a small dragon in the distance. The McTaggarts hope it will be declared part ► of a "quiet zone" for an astronomical array in the area.

The mountain may have a spiritual heritage as well. In the past, geologists have found rocks placed in a circle on its summit. At one time, Aborigines may have used the mountain to initiate their young people, giving them a deeper understanding of their relationship to the land.

Carol firmly believes that the geologic treasures at Mt. Narryer belong to humanity and need to be protected from mining operations. "I've got a list of white knights to help me," she says, "geologists who are willing to support the cause. And Sandy critiques all the letters I write to politicians."

A National and World Treasure

The next day we reluctantly leave for Shark Bay on the coast and then the long flight home. But we keep in touch with Carol, who writes to tell us that after several years, her efforts have finally paid off. In 2009, the government granted National Heritage status to Mt. Narryer, giving it limited protection as a scientific site. Her ultimate goal is to gain World Heritage status for the mountain through the office of the United Nations.

The protection comes at a time when large areas of Western Australia are being mined for iron ore, nickel, and rare earth minerals. Jack Hills, site of the oldest zircons ever found, has been largely destroyed for its iron deposits, with only a small section set aside for geologic study.

We learn that Mt. Narryer could gain permanent protection if a large astronomical array is built nearby—one square kilometer of radio dishes listening to the universe. The mountain would be part of a "quiet zone" around the array, established to prevent nearby mining operations, ranching, or local airplane flights from affecting sensitive radio equipment. The mountain would make a fitting guardian for astronomers studying the origins of the cosmos.

Someday, Carol knows, she and Sandy will move away from their land. As keepers of Mt. Narryer, they want to know that the mountain they love will endure as it has for millions of years—guarding some of the oldest fragments of Earth.

Nuuk

AKILIA ISLAND

63° 55′ 59″ N
51° 40′ 01″ W

Atlantic Ocean

Iceland

AKILIA ISLAND

GREENLAND

The slow journey from pole to pole ...

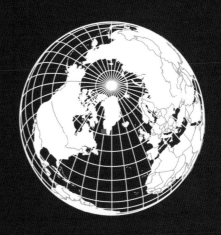

AKILIA

Akilia is in a cluster of islands on the west side of Greenland. It is part of the world's oldest oceanic crust.

*A*kilia Island is strikingly different from the deep isolation of Mt. Narryer in the Australian Outback. From crowded Nuuk, the capital of Greenland, we take a spine-jarring boat ride across 15 miles of choppy sea dotted with small islands and fishing boats.

Arriving at Akilia, we scramble off the boat and crawl on our hands and knees like children over huge granite boulders to reach the island itself. Later we learn that crawling onto an island is the Inuit way to show respect for its spirits. That accidental courtesy will serve us well here.

Akilia allows us to see and sense the restless forces that shaped its 3.8-billion-year-old stone. As Earth's massive plates pushed this land northward from the south pole, the stone softened, split, and folded, creating the vivid patterns and formations we find on the island.

ISLAND
GREENLAND

AKILIA
ISLAND

When these rocks split, lighter minerals flowed into the cracks, preserving Akilia's history in stone.

This power connects us to the turbulent violence and creative

fire that are constantly bending and shaping us as artists.

Akilia teaches that out of great pressure and change, great beauty can emerge.

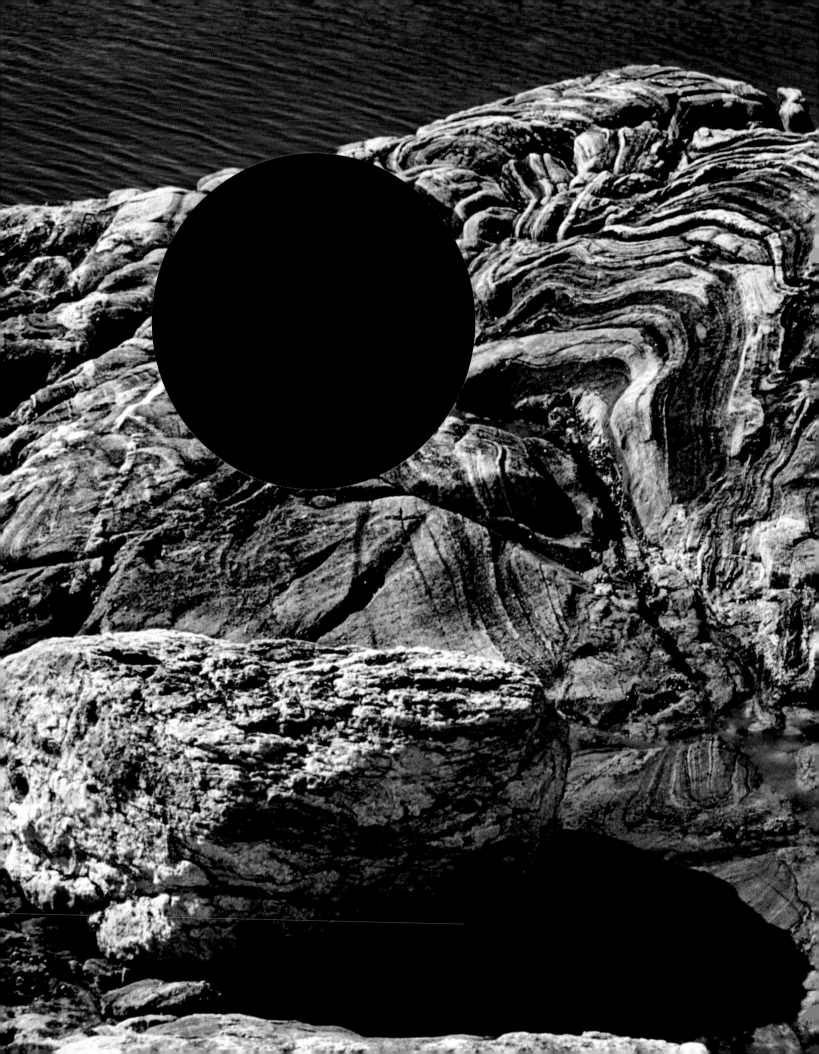

"MEET YOU IN NUUK!"

*O*ur first advice about traveling

to Akilia Island comes from Dr. Stephen Mojzsis at the University of Colorado in Denver. He has spent years studying the rocks of Akilia.

"You'll have to hire a boat to take you out to the island, but it's a scientific sanctuary, so everyone knows where it is. The place is very Tolkien-like, only about a half-mile wide. You're going to see some of the oldest oceanic rock on the planet. And you'll have 24 hours of daylight to photograph it, because in July the sun never sets!"

A lot has changed since our journey to Australia. Lynn has moved to Switzerland, and Sue has taken

a job with a new publishing house. But deeper changes, more difficult to articulate, are also at work in us. The power we felt at Mt. Narryer continues to sink into our lives as if we had never left the Outback. We come to Greenland with less of the "city" in us, already able to see and hear more of what lies beneath the surface.

We arrive in Nuuk to find the entire west coast of Greenland and its islands shrouded in dense fog; even the massive glacier covering most of the mainland is barely visible. This gray light will make our work more difficult, but fog is not the only challenge we face.

A Land Born of Fire

We're not prepared for the explosion of activity that marks the subarctic during summer. Twenty-four hours of daylight means 24 hours of traffic in Nuuk, fishing boats leaving shore at all hours, and cargo hauled nonstop from an endless stream of freighters in the harbor. Even natural rhythms fail us as we watch the sun dip low to the horizon only to rise again and circle overhead. We won't see night for the entire week we're here.

And Nuuk seems to be a gathering for the peoples of the world. We hear Swedish, Danish, German, English, Chinese, Inuit, Norwegian, even Romanian and Bulgarian swirling around us. We find refuge from the chaos by visiting the local university's exhibit about Akilia and Isua, another ancient site several miles inland. At Isua, geologists have found the oldest traces of life, dated at 3.8 billion years ago. This exhibit sparks the idea of a connection among ancient stone, ancient life, and humans, an idea that will take another two years to mature in our work.

The exhibit shows that icebound Greenland was originally formed by magma from Earth's mantle and by sediments laid down on the ocean floor that accumulated slowly until thrust above the surface. As Earth's moving continents pushed Greenland from the South Pole toward the Arctic, collisions with other landmasses created immense heat and pressure that twisted and transformed Akilia's rock. The harsh Arctic weather has also attacked the stone, making it difficult for geologists to trace

From late May to late July, the sun never dips lower than the horizon and then rises again. This long day means that people won't see the moon, the stars, or the aurora borealis for three months.

its origins and history. We wonder what the voice of the land will sound like.

"The spirits have found your project worthy."

The fog stubbornly refuses to lift, and we have only two days left. We stick to our plan and hire a boat from the Tourism Office to take us out to Akilia the next day. That evening we're introduced to Bjarne Kreuztmann, the former mayor of Nuuk who has just returned from abroad. He takes us out to dinner at Nuuk's best hotel. As a jewelry maker, he finds our rock project intriguing. "Don't worry about the weather," he jokes. "I'm part Inuit and was brought up in the old ways, so the spirits know me. The weather will clear because I'm back home."

Within the hour, we watch in amazement as the thick fog starts to pull back like a veil lifting. The midnight sun breaks through, casting a deep golden sheen over the bay and the city around us. Bjarne smiles and lifts his glass in salute. "Apparently, the spirits have found your project worthy." The next morning, the sky remains brilliantly clear for the work we came here to do.

*W*e crawl over the last boulder and stand up

to get our first glimpse of Akilia Island. A line of snow-covered hills, on a larger island nearby, seems to shelter tiny Akilia. The light dazzles us, illuminating every stone, every plant with startling clarity. Bright red and orange seaweed cling to the rocks while lichen and summer flowers flourish even in the cold wind blowing off the ocean. We smell briny seaweed and the island's damp earth.

We quickly find the site we came to photograph (circled at right), an area geologists have cleared of lichen. If the deep calm at Mt. Narryer had steadied us, Akilia overwhelms us with its stunning formations. Our newborn confidence is momentarily shaken. Do we have the artistic skills to capture these stories?

Yet each time we frame the sinuous stone in our lenses, something pushes us to try another angle, move a few steps to the right or left. We are being challenged by a strong, clear voice: "How can you doubt the force that moves in you? Look what it can do with stone!"

ISLAND OF SPIRIT AND STONE

*W*e are not alone as we work.
Ghostly footsteps and whispers follow us around the site. Once or twice Sue hears Lynn walking behind her, only to turn and find no one there. Perhaps our gesture of respect encourages the spirits to draw close. But if they are trying to tell us something, it's in a language we don't yet understand. In time, the spirits fade, and we hear only the wind again.

The wind picks up, tearing the heat away from our bodies until we pull on more jackets to ward off the cold. Our bare hands grow numb as we work the cameras and change rolls of film. Every time we take shelter behind a few boulders, thirsty Arctic mosquitoes attack and drive us back into the open. But what we see through our lenses is worth any hardship.

The power here not only shaped stone into sweeping patterns but also worked smaller, nuanced designs into the surface. Our own creative eye is quickened, allowing us to see designs we didn't notice before.

When everything can speak, true knowledge appears.
—Inuit saying

Lynn (left) and Sue (right) bundled against the Arctic winds on Akilia.

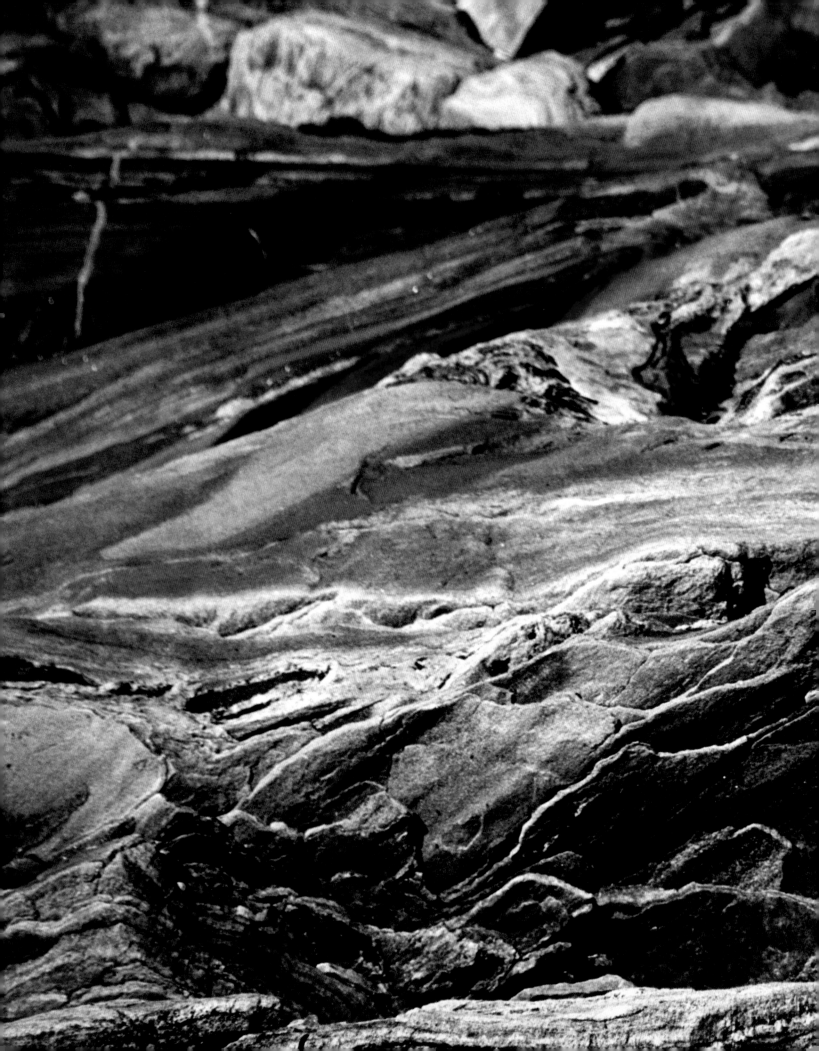

Rugged lichen and small, bright flowers draw our attention now. We don't know the names of any of the plants; it's enough that they belong here.

As we move closer to each one, there's a strange feeling of recognition, as if in some child-like way we know each other. Maybe the spirits are trying once again to give us knowledge. Then the feeling passes, and we focus instead on how tenaciously these delicate plants hug the ground or how firmly they attach themselves to rock faces to weather the storms that lash Akilia.

The paradox of such softness and stone never fails to intrigue us. In this radiant light, even pools of water can change into gold.

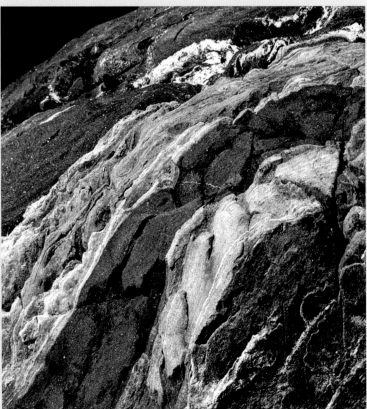

The sandstone formations on Akilia (upper left) look like ruined fortresses near a wall of metamorphic stone (upper right).

A field of granite boulders (above) are dissolving grain by grain under thick mats of dark lichen.

This is an island that Tolkien

might have imagined—Dr. Stephen Mozjsis was right. The spirits seemed to have divided Akilia among themselves to fashion their own unique, sometimes bizarre, landscapes.

We stroll among sculpted pillars of sandstone, thrust skyward like the ruins of old fortresses. Only a few yards away, a waterfall of quartzite, hornblende, and granite cascades over a small rise. At the far end of the island, we discover a field of granite boulders that resemble a herd of primitive animals resting in the grass. No wonder the spirits are so protective of Akilia.

Whatever force is at work here appears to transform everything it touches. We have felt it molding our own artistic sensibilities to strengthen what is emerging in our lives. It is on Akilia that we begin to inhabit our work as artists.

The next day on the flight leaving Greenland, we see the same transformation pushing deeply, inexorably in the mainland as well. Each year Greenland's glaciers recede farther, revealing new coast land that has been buried under ice for thousands of years. The Arctic is melting.

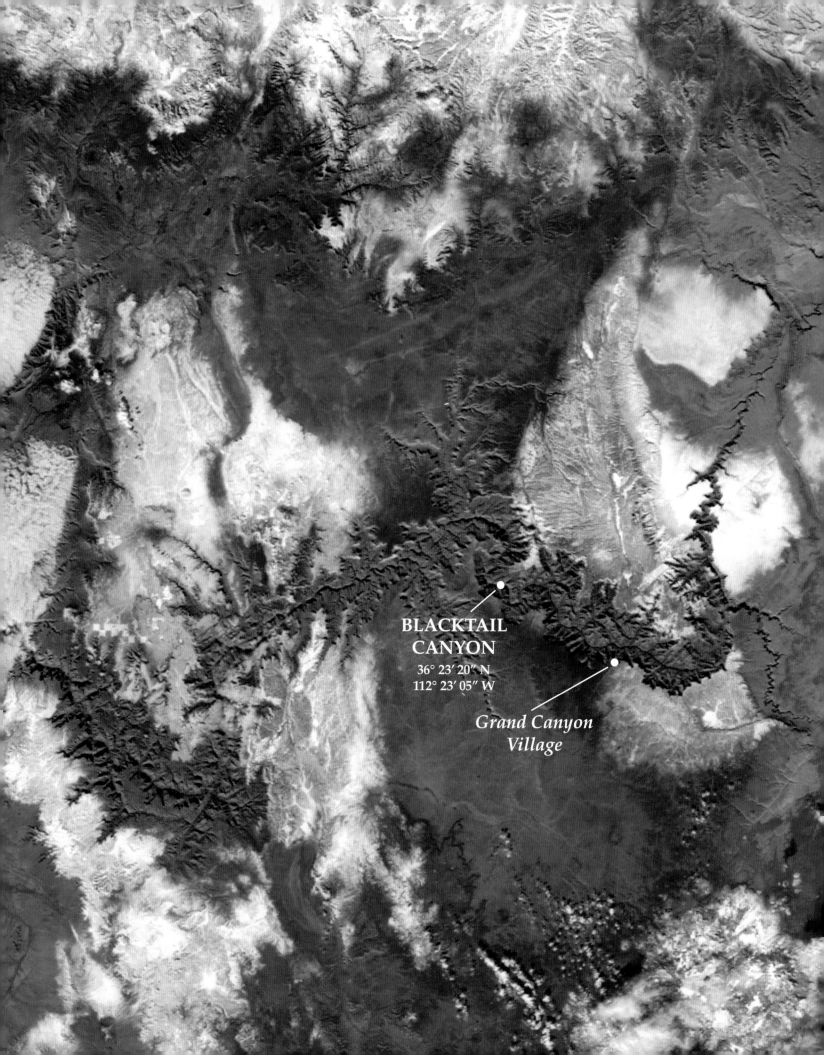

**BLACKTAIL
CANYON**

36° 23′ 20″ N
112° 23′ 05″ W

*Grand Canyon
Village*

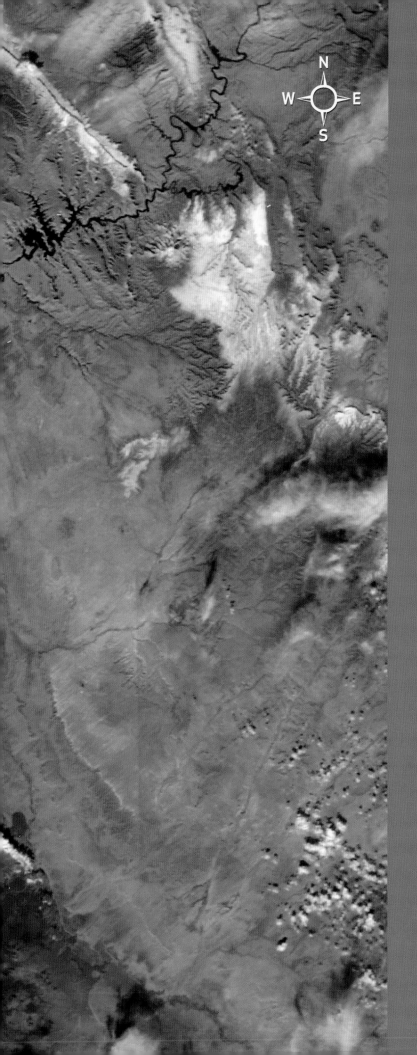

BLACKTAIL CANYON

GRAND CANYON, UNITED STATES

A billion years of Earth history is missing ...

BLACKTAIL

For the only time in our travels, we will not make a journey together. Lynn must return to Switzerland, so Sue continues on, joining a group led by outfitters to raft the Colorado River. The goal is to reach Blacktail Canyon, where the river has exposed stone 1.7 billion years old. Only in Blacktail can we actually touch its ancient surface.

After three days of rafting the rapids, our group finally reaches Blacktail at mile 120. We follow the canyon's narrow passage into a chamber whose basement layer is a dense, ancient stone called Vishnu schist. The rock seems bowed under the weight of the continent resting above it.

But these walls speak of loss as well as strength. The rock layers reveal that nearly one billion years of Earth's history is missing—eroded away by rains and scouring winds. With no fossil traces left behind, we will never know what lived and died here throughout all that time.

The canyon's walls are a

haunting reminder to us:

RAFTING THE COLORADO

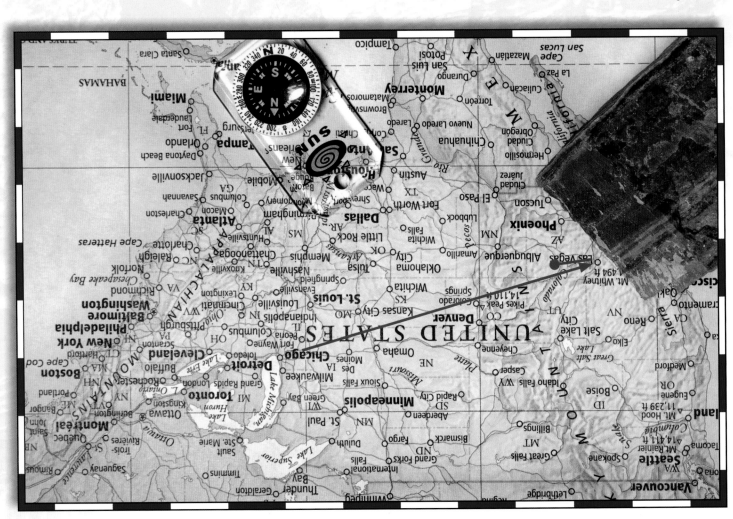

We hear the throaty roar of

Waltenberg Rapid long before we see it. The sound slowly swells as we drift nearer and see the foaming, churning water dead ahead. Our river guide grips his oars and yells, "Everybody hang on!"

The raft plunges into the first wave, and half the river seems to explode over the bow, drenching everyone on board. Another wave rears high above us, but we don't raft over it, *we shoot through it.* After thirty seconds of screaming terror, we're suddenly out of the rapid and drifting quietly downstream. We start bailing out the raft in the July heat—a scorching 115 degrees.

Another Way of Traveling

In three days, we've rafted 23 miles from Phantom Ranch, where the expedition began. Blacktail Canyon lies only eight miles farther downriver, but we'll have just a few hours to photograph the site, if the weather allows it.

Last week, powerful thunderstorms sent flash floods surging through Blacktail and left the Vishnu schist under 12 feet of water. This morning, we woke up in camp to a cloudy dawn; now, rumbles of thunder echo off the canyon walls, and our guide keeps nervously checking the sky.

On most whitewater rivers, rapids are rated 1 to 5, with 5 the most dangerous. On the Colorado River, the rapids are rated 1 to 10, but the river is so deep that all rapids can be navigated. Hermit Rapid, shown here, is rated a 9.

The journeys to Mt. Narryer and Akilia have taught us that we always travel at the indulgence of forces more powerful than ourselves. It seems natural to ask their permission before approaching a site and to listen carefully for the answer.

Now the request is for safe passage to Blacktail and for good weather to document the stone. Surprisingly, it's the river, not the stone, that seems to speak for this place, quietly giving an answer: The weather will hold until the group leaves Blacktail Canyon.

A Vertical History

Even by mid-morning, hot winds are sucking the moisture out of us, and we keep dipping our shirts in the river to stay cool. Maidenhair ferns struggle to survive along the cracked shoreline, and the few mountain sheep climbing down to drink at the river look thin and stressed. We raft through a section of the canyon where the Vishnu schist, named after the Vishnu Temple formation, can first be seen in the rock walls.

The river guide points out that the canyon's towering cliffs represent a vertical history of the land, with the youngest layers at the top. As the Colorado plateau slowly lifted, the river cut deeper through rock from 100 million to 600 million years old until finally reaching the Vishnu schist. As we float by the rock walls, the guide shows us where a seam of older rock is cutting through younger layers, as if they were shuffled by a whimsical hand.

The schist itself formed nearly two billion years ago when two landmasses merged. The crushing pressure destroyed most of the original stone and fused what was left with other minerals to create a denser, tougher rock. This ancient layer has witnessed the slow formation of North America, including times when inland seas cut the continent in two and laid down sediments nearly a mile thick.

We drift past an area where the dark schist extends into the water. Its presence anchored this land long before the river and its canyon even existed.

ENTERING THE CANYON

We pull our rafts onto a sandy beach where the Blacktail trailhead begins and hike through the stifling heat until we reach the canyon's narrow entrance. We're not prepared for the sheer height and rugged beauty of the canyon walls.

Over millennia, winds and flash floods have carved deep lines in the sandstone and etched honeycomb patterns in softer limestone. When shafts of sunlight penetrate, the canyon seems lit from within.

There couldn't be a more fitting place to house the Vishnu stone. The power emanating beneath our feet, though not as old as at Mt. Narryer, feels as undisturbed and welcoming. The deeper we walk into the canyon, the more the power flows into us.

At one point, our guide stops and in the silence begins to read the words of explorer John Wesley Powell, who loved the canyon. But the sounds that resonate back from the walls are older, like the murmur of a river.

Above We tied up at the beach after navigating Blacktail Rapid and spend the morning in the canyon. **Near right** The tiny figure at the bottom of the picture shows the scale of the canyon walls.

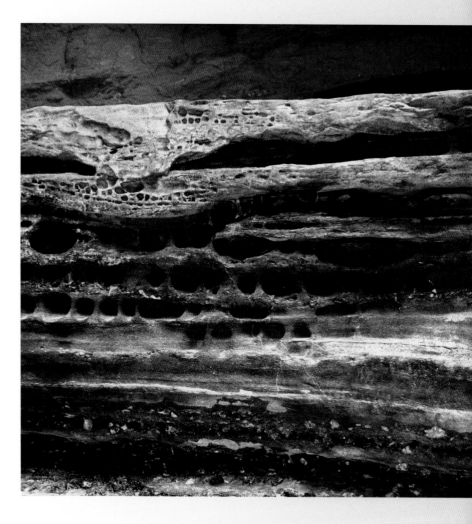

*A*t last we enter a place where the Vishnu schist forms a chamber hidden from the sun. The stone surface bears long scars where heat and pressure softened and split the rock, allowing white quartzite to fill the gaps. The schist feels cool to the touch and has the same dry, acrid smell as the granites of Mt. Narryer and Akilia. And there is the same eerie sense of recognition that seems to vibrate in the bones.

The light is difficult, but the weather has held long enough for us to photograph the complex lines and patterns marking the surface of the schist. Someone finds the fragment of a fossil on the chamber floor, and others find more sea fossils in the younger sandstone. In one place, the rock seems to rise out of the chamber floor like the back of a giant animal whose species has long since disappeared from the continent above us.

<blockquote>
❝ *Maybe an artist's true work is going after what is lost.* ❞
</blockquote>

*T*he guide tells us there is an even more remarkable formation in the chamber called the Great Unconformity. He shows us a wall where a younger sandstone layer rests directly above the far older schist. "The Great Unconformity shows that a billion years of Earth history has eroded away," he explains.

Most of the group moves on, but a few of us linger, talking quietly about the gaps in our lives and family histories in which stories have been erased by tragedy or by deliberate forgetting. Some of us recall years when we stopped creating—all the stories and artwork eroded away. We wonder if such losses are irretrievable or if something persists after all that time. Maybe an artist's true work is going after what is lost.

Suddenly a faint sigh, like the last breath leaving a body, moves past us as if some primordial life is still imprinted in this chamber. We hold our breath for a tense moment and listen.

"Back to the rafts!" the guide shouts, and the moment is gone. Perhaps such places hold memory in more ways than in stone or in tiny crystals. Or perhaps we felt only a gust of wind. Still, as we walk back to the river, something of the memory and voice of this canyon follows us.

<blockquote>
<blockquote>
<blockquote>
<blockquote>
79
</blockquote>
</blockquote>
</blockquote>
</blockquote>

ACASTA RIVER
65° 10′ 30″ N
115° 30′ 30″ W

Yellowknife

Great Slave Lake

ACASTA RIVER

NORTHWEST TERRITORIES, CANADA

On a remote island in the tundra ...

ACASTA RIVER

NORTHWEST TERRITORIES, CANADA

*O*ur floatplane pitches and rolls as we skim low over the immense tundra of northwest Canada. We clutch our stomachs while the pilot homes in on a small, unnamed island on Acasta River. Here the oldest skin of Earth, four-billion-year-old Acasta gneiss, lies exposed to view.

Our pilot finally eases the plane down on the choppy water and taxis toward the site. We're dismayed by the steady mist, but it enhances the green and blue-black tones of the gneiss, giving it the look of whale skin. We imagine the huge mammal surfacing for air from the depths of the Canadian Shield, the bedrock of North America.

Earth's memory is intact here; this greenstone has witnessed the formation and drift of continents and the entire rise and evolution of life. Deep Time is captured in feldspar, apatite, olivine—the poetry of stone that tells us this skin has survived all of Earth's cataclysmic changes and still nurtures life.

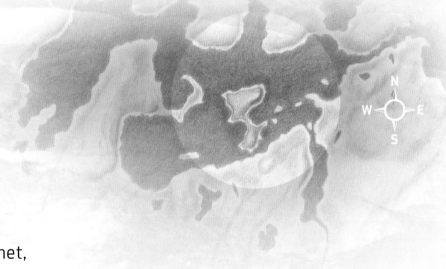

Our culture claims we exert

dominion over the planet,

but the stone tells another story:

we are utterly dependent on Earth for our survival.

"TREASURE ONLY A GEOLOGIST COULD LOVE"

"*If you're going to Acasta River*

island, you should talk to Walt Humphries and Diane Baldwin, " the Yellowknife tourist agent tells us. "He's a mineralogist and she's a district geologist. They know all about the island."

Yellowknife, the capital of Northwest Territories in Canada, sits on the edge of the Great Slave Lake surrounded by a sea of tundra as vast as the Outback. We've arrived in midsummer to find Yellowknife caught up in the frenzy of a major diamond strike. Every day at 9 a.m. underground blasting at the diamond mines shakes the city like

a small earthquake. Mining companies have carved roads through the surrounding forest and left mounds of rubble blocking streams and wetlands. Nearby, an abandoned gold mine rusts in summer rains, its land barren as the moon.

What Lies Beneath

Yet despite the damage, we are amazed to feel the clear, peaceful presence of Deep Time beneath this land. Only at Mt. Narryer have we encountered anything close to it. Maybe Walt and Diane can help us understand the paradox we find here.

As Walt welcomes us into his home, one of the first things he shows us is a geophysical map of North America on the dining room wall. "The Canadian Shield isn't just under the Northwest Territories," he says. We see that the shield extends underneath most of Canada and into Michigan. Diane adds, "This stone is a survivor of the Archean Eon, four billion years ago. Shields like this are under every continent on Earth except Antarctica."

We're beginning to understand where this voice comes from and why we are encountering it at each site. Like whale songs criss-crossing the ocean, the echoes of the shield must reverberate through the world's skin.

Walter says, "The shield has gold, silver, rubies, even diamonds. But the island where you're going has a treasure only a geologist could love—the Acasta gneiss. Gneiss is the hardest, densest form a rock can be changed into—sort of the end of the line."

Diane hands us a geology article. "When the rock was dated, no one could believe it. How could a shield that old still be intact? Geologists ran the dating three times before publishing their work."

The clouds briefly cleared for take off and then closed in again during our rough flight. The plane's cockpit was about the size of a Volkswagen.

The Flight North

Walt and Diane give us the GPS coordinates for the island and wish us luck. The weather hasn't been good all week, and the site lies 150 miles almost straight north by floatplane.

That night we ask permission of this place to visit the island and photograph the stone, and then add a plea for the weather to clear. The answer comes back: you can go to the island, but the weather will not change. We think of Bjarne in Greenland. Have the spirits lost interest in our work?

Finally, with only two days left, we give up waiting and hire a floatplane to take us north, almost to the Arctic Circle. The clouds part briefly to show the tundra spreading below for thousands of square miles. The plane carries its own fuel supply. There is no fuel, food, or shelter where we are going.

*T*he pilot ties up the plane on the beach, and we
struggle unsteadily to get out of the cockpit. The minute our grateful feet
touch the sand, we spot the site (circled, and at right), a small island of stone
surrounded by black willow brush. Even from a distance, it radiates power.

We'll have to work fast before we lose the light. But when we step onto the
stone, we forget time and weather. The stone seems to pulse with a living force
as thick ropes of quartz insinuate their way through its dark layers. Long, thin
scratches reveal where glaciers once ground across the surface.

The knowledge carried in this stone is staggering. It dwarfs everything we
learned or sensed at the other sites and yet includes it as well, like a sage
putting all the lessons together. We stop thinking; we take picture after picture
as the voice of this place keeps sounding inside us.

Top We were drawn to the Zen-like markings on the stones, created by the same immense forces that can melt solid rock.

Bottom Glaciers forced these layers together as if experimenting with contrasts in color and texture.

We focus on smaller, more detailed formations and find other images from the stone flowing behind our eyes—scenes of inland seas, savannas, immense sheets of ice grinding and compressing the rock into the shapes and designs we see.

Black flies and mosquitoes swarm the area but stay out of our camera lenses and, strangely, fail to bite. They bounce off our jackets as if we were just another part of the island.

$\mathcal{M}$*ist turns to drizzle, and the* greenstone acquires a fine sheen that looks more than ever like skin. Now it's clear why we were told the weather wouldn't change. As we photograph quartz ribbons falling like water through the stone, we realize the spirits still find our project worthy.

This island of stone taps directly into the shield below and, perhaps, into the network of shields around the Earth. We look at the island with eyes that are seeing more from moment to moment. The shrubs, brambles, and stone weave around us until all separation dissolves. We are not going into the wilderness and returning to civilization. We are enfolded in a dazzling, sometimes dangerous Eden we have never left.

▲ LIFT UP

Lift Up Carefully

Lichen create endlessly varied patterns from one stone to another. Like the best abstract artists, their compositions show a flare for subtle hues, complex design, and dramatic contrasts. The bottom photograph is a close-up of the boulder shown at top right.

Patterns of lichen cover the rock faces, resembling maps of an earlier world. The lichen smell salty as the ocean. Their work is infinitely slow, patient, inevitable.

There are no spirits whispering to us on Acasta island, yet we seem able to perceive more as we move quietly among the stones. Silence draws things together.

Now we can see that lichen and stone are linked in an age-old cycle. Over millennia, lichen breaks rock down into soil; larger plants take root and begin to grow, and slowly a once-barren land— like the young Earth—becomes filled with life.

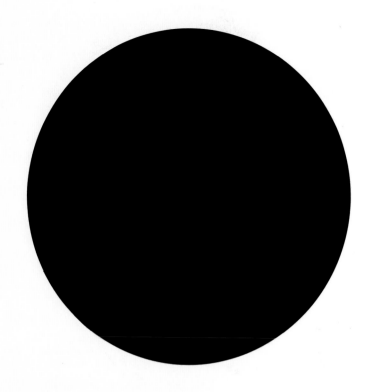

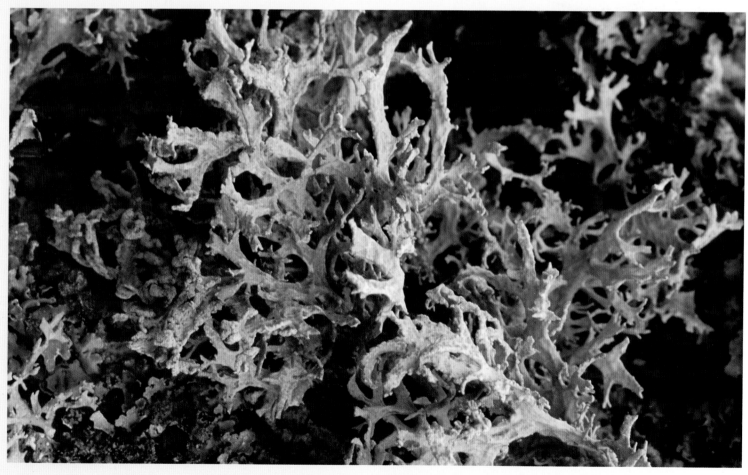

In this close-up taken by researcher David Aubrey, lichen tendrils appear like a miniature forest on the stone. These delicate-looking colonies can survive scorching summer heat and winter temperatures of more than 100 degrees below zero.

> **" We're carrying Earth's oldest story . . . ""**

The last patches of faint sunlight disappear as darkening clouds descend over the site. The worried pilot calls us back to the plane.

We hike down to the beach and squeeze ourselves into the cockpit. Before taking off, we ask the pilot to taxi by a bluff on the far end of the island where a solid wall of Acasta gneiss drops 30 feet into the brackish water. The rock face is too slick and dangerous to climb, but we're able to get good shots of the gneiss as we pass by.

The plane picks up speed and lifts off; Acasta island is quickly lost in a patchwork of rivers, lakes, and islands below. We feel the power of the stone vibrating in us still; we're carrying Earth's oldest story, and hope we can give voice to it in our work. But before we leave Yellowknife, this ancient land has something more to show us.

Yellowknife

GREAT SLAVE LAKE
61° 40′ 00″ N
114° 00′ 00″ W

GREAT SLAVE LAKE

NORTHWEST TERRITORIES, CANADA

Life flourished in an ancient shallow sea ...

GREAT SLAVE LAKE
NORTHWEST TERRITORIES, CANADA

*W*e *find the fossil slab that Walt Humphries* told us about in a boat captain's backyard in Yellowknife. The fossil, taken from Blanchet Island in Great Slave Lake, is the remnant of a vast colony built by ancient cyanobacteria 1.7 billion years ago. These early life forms thrived along the shores of a warm inland sea that once covered the middle of North America.

Studying the fossil closely, we're mesmerized by patterns that resemble miniature ravines, moraines, dry riverbeds, and rippling hills that rise and fall across its surface. We don't realize yet that the cyanobacteria builders are the greatest terra-formers Earth has ever known. Over nearly two billion years, they helped shape the world that sustains us even now.

Yellowknife

GREAT SLAVE LAKE

This close-up of the fossil shows the complex structures and layers built by bacteria.

This fossil, far more than a mere record

of the past, reveals that

All life shares a history we are

only beginning to learn.

The bacteria
lived only in the
top layer of the
colony. The central
core became as hard
as concrete.

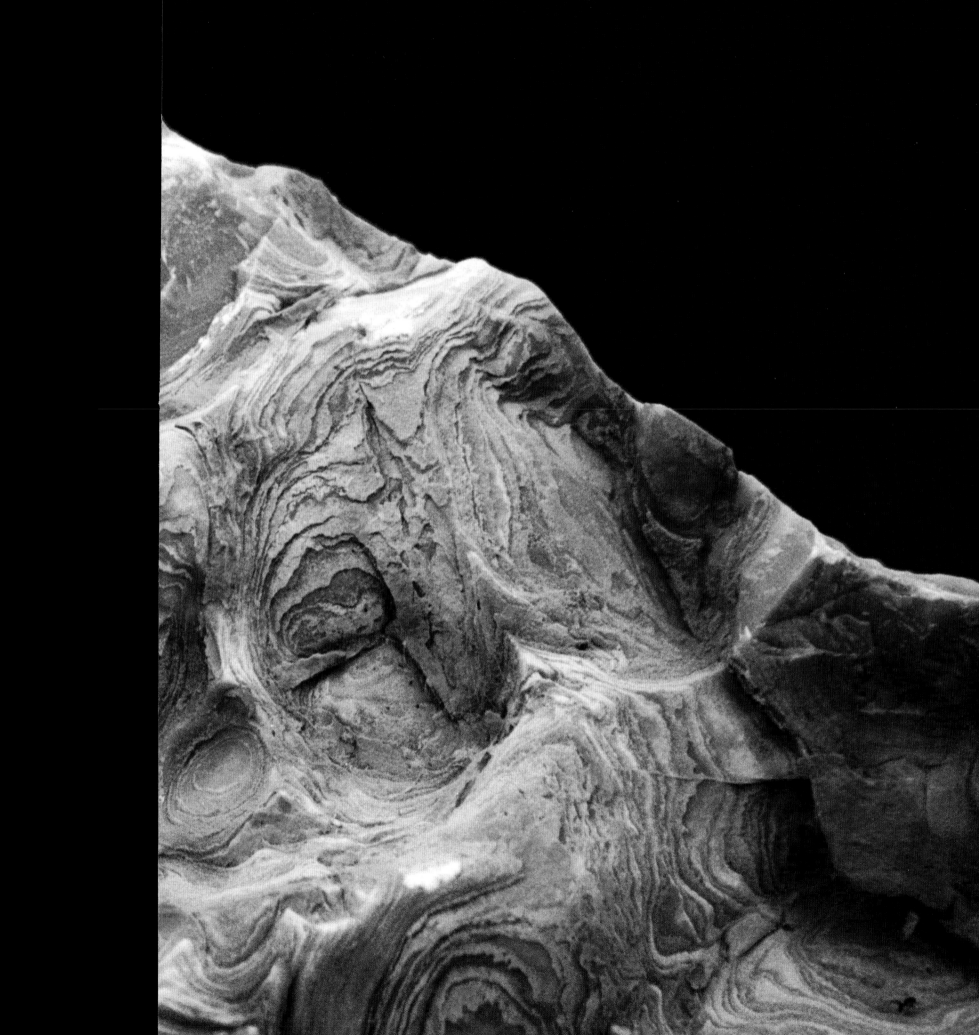

THE FOSSIL IN CAPTAIN SMITH'S BACKYARD

The road leading us to the fossil

begins after we return from the Acasta River island. With only one day left in Yellowknife, we have a chance to ask, "How did Great Slave Lake get its name?" To our surprise, no one knows. Instead, they tell us this body of water is the tenth largest lake in the world, is more than 800 feet deep, and offers some of the best fishing in Canada.

We discover that not even Walt Humphries, the mineralogist who helped us get to Acasta River, knows the answer. "Can't say I've ever heard what the name means, but do you know the lake has fossil stromatolites?" It turns out these fossil

stromatolites—stony colonies built by ancient bacteria—are abundant along the shoreline of the lake's Blanchet Island.

Walt continues. "A boat captain I know, Dave Smith, has a good-sized fossil in his backyard that you could photograph." We write down the address but are no wiser about the origin of the lake's name.

Story of the Lake

We've learned since Greenland that if local white people can't tell you about place names, other locals can—in this case, an Inuit group called the Dene who

The Dene tepee and the raven symbol painted on this granite boulder in Yellowknife honor the Inuit culture. Yellowknife includes the number of ravens in the city's population count.

have lived in the area for several thousand years. The spirits still seem to be with us, because that afternoon we happen to meet Tim O'Loan, a Dene who is the chief aboriginal negotiator for the Inuit in the Northwest Territories.

He invites us up to his office, and when he hears our question, his solemn face breaks into a broad smile. "Of course I know what the name means. My uncle told me the story." In earlier days, he says, most Inuit followed the caribou herds as they migrated to the Arctic Circle and back. But one Inuit group settled around the lake and found they could get nearly all the food they needed from its abundant waters. Why bother to follow the caribou?

Over time, the other Inuit began calling them "slaves of the lake." When the French and English arrived, they thought the term referred to the lake itself and called it Great Slave Lake. We laugh. White people—we're always getting the details wrong.

Fossil from the Island

After leaving Tim's office, we drive through the winding streets of Yellowknife until we locate Captain Dave Smith's home. No one answers the door, but the back yard faces an alleyway, so it's easy to spot the three-foot fossil near the property fence. The stromatolite domes and layers are surprisingly well preserved.

The captain has planted the slab solidly in the ground and enclosed it in a circle of stones, as if he meant to set a protective ring around it. The dim, cloudy afternoon provides a dark background that will make the slab easier to photograph.

We step inside the rock circle and kneel down close to the fossil. All we want are a few images of the brightly colored lichen and close-ups of the "landscapes" we see. The fossil, however, gives us a great deal more.

Slabs like this one are cut from fossil beds found on the eastern side of Blanchet Island in Great Slave Lake.

When we return home and

develop the images, we get an inkling that something unusual happened in the captain's backyard.

At first we're struck by the sheer artistry of the lichen patterns tracing the fossil's delicate, rimmed lines. Remarkably, this lichen is partly made of cyanobacteria. In an ironic twist, it seems the ancient bacteria built a structure that two billion years later their descendants are slowly turning back into sand.

But as we look at the images, a living presence keeps drawing us in deeper, asking us to see it . . . and we can't. *What are we missing?*

The photographs continue to baffle us like tantalizing puzzles. According to the history of the ancient cyanobacteria, they first appeared on Earth more than 3.5 billion years ago. In that eon of Deep Time, the young planet lacked oxygen. Scientists believe that methane- and sulfur-breathing bacteria tinged the oceans a greenish hue while the thick atmosphere flared a reddish-orange.

For three quarters of Earth's history, these bacteria ruled the planet, using sunlight as fuel to build massive colonies. Over time, they released enough oxygen into the oceans and air to transform Earth into a blue world. The oxygen we breathe is a legacy of their work.

Now when we spread out the images, the fossil's maze of whorls, ripples, and concentric patterns seem more vivid, as if they still pulsed with life. We feel a compelling urge to touch that bright, hidden energy with our hands—to draw, work in clay, *do something* to express the essence of it.

> *Really old things in nature give you the most power.*
> —*George Blondin, Dene elder*

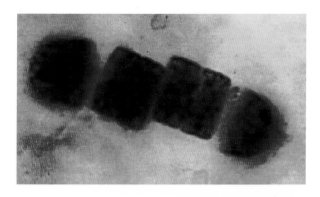

Top Close-up of ancient cyanobacteria in a fossil trace.
Middle Bubbles of oxygen are still produced by modern cyanobacteria in ponds, lakes, and slow-moving rivers.
Bottom Earth as it might have looked two billion years ago, when its atmosphere and oceans lacked oxygen.

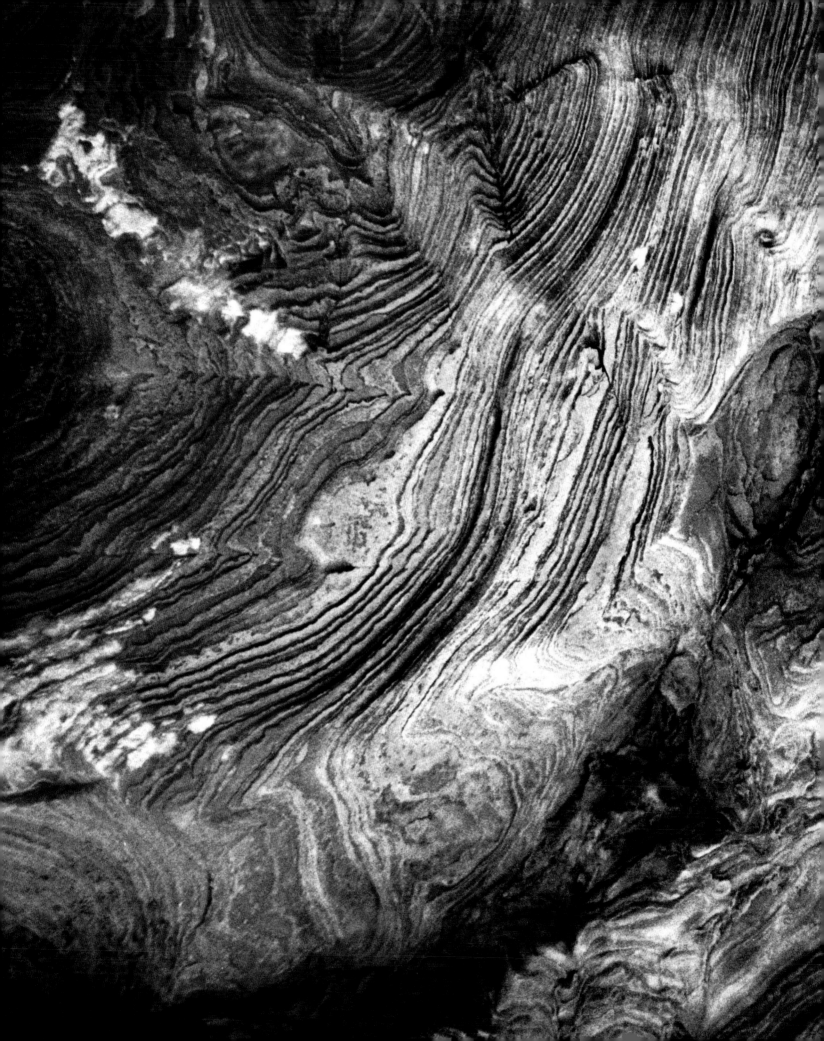

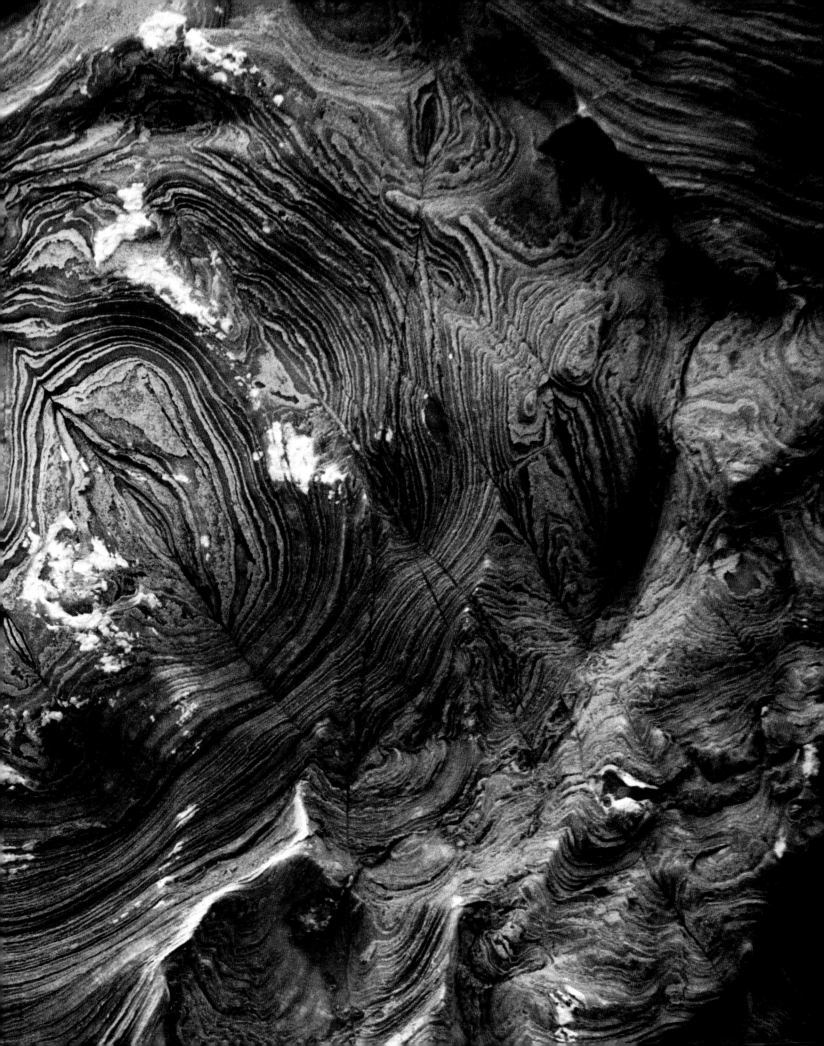

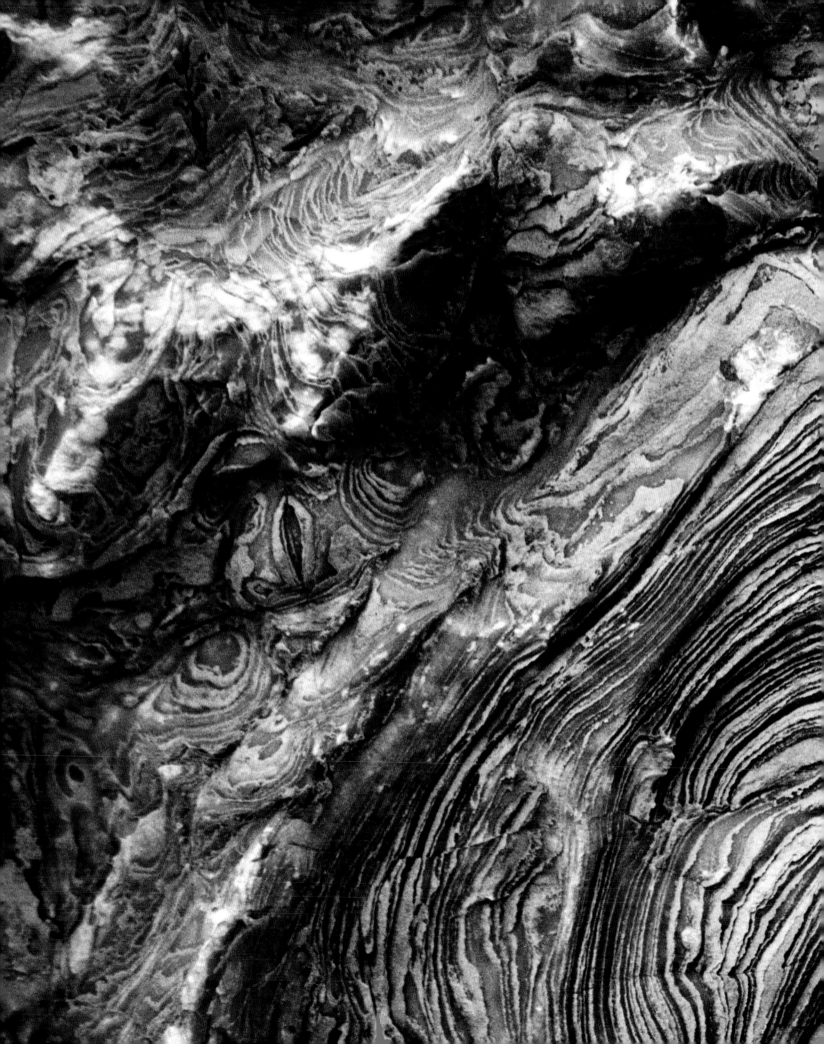

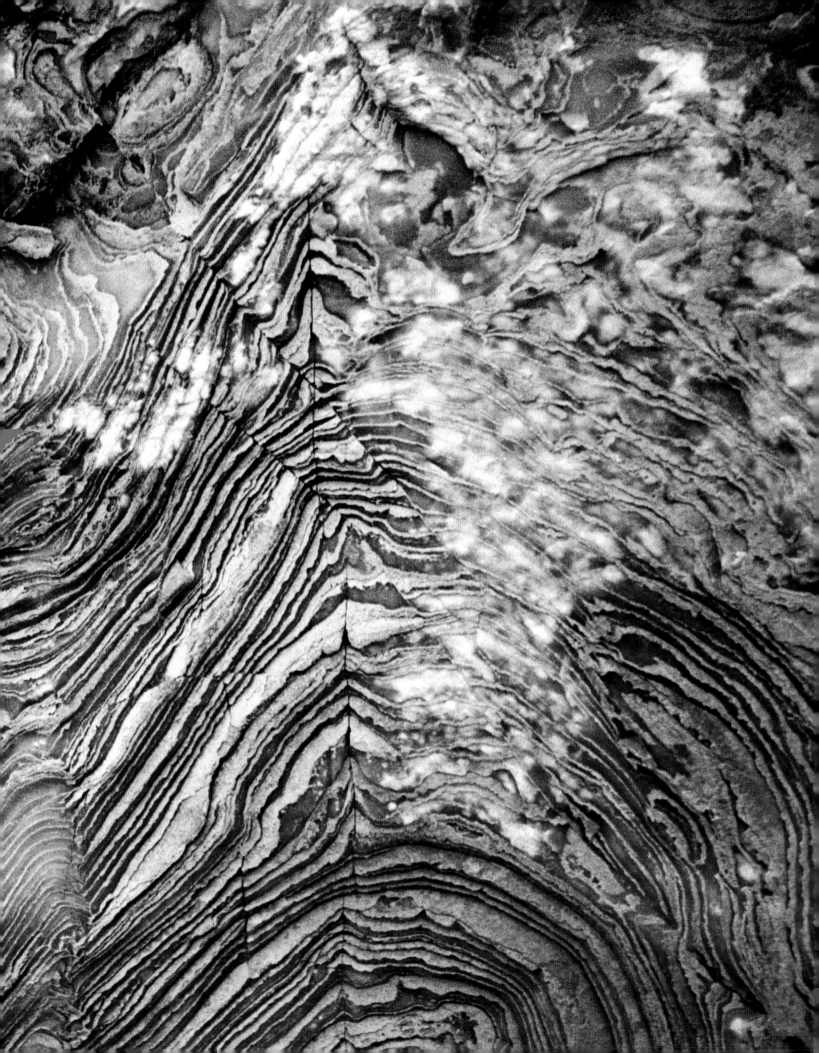

SHARK BAY
25° 55′ 60″ S
113° 32′ 32″ E

● *Hamelin Pool*

Indian Ocean

SHARK BAY

WESTERN AUSTRALIA

Living descendents of early life ...

Mt. Narryer
▲

SHARK BAY
WESTERN AUSTRALIA

We hurry to take our last pictures before the morning tide completely submerges the stromatolites at Hamelin Pool in Shark Bay. The tidal waters, ten times saltier than the Indian Ocean, are already lapping at our boots.

We have traveled here to visit one of the last colonies of living stromatolites in the world. They are all that remain of the huge, reef-like communities once found along every shoreline on Earth.

At first, these light-gray to nearly black stromatolites appear lifeless. Yet the top layers are teaming with colonies of bacteria whose lineage stretches back into Deep Time and whose forbearers helped build the world we know.

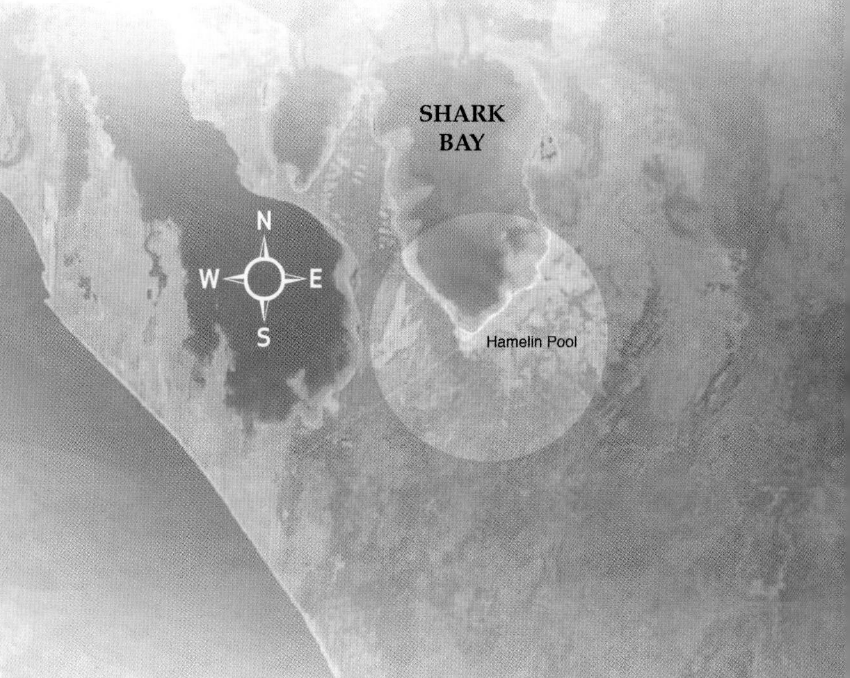

These humble survivors are not

only among the most

enduring of species,

*in many ways, they are
the ancestors of us all.*

SHARK
BAY

Hamelin Pool

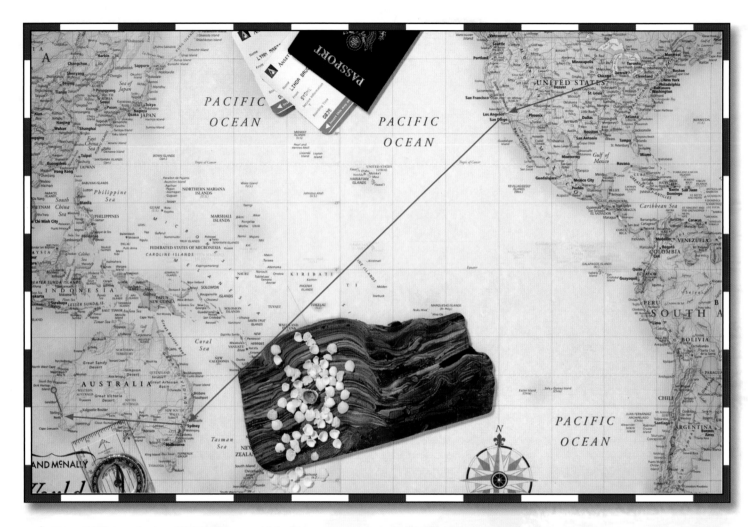

[NOTE: We traveled to Shark Bay and Mt. Narryer on the same trip. But after photographing the fossil stromatolite at Great Slave Lake, we realized that the story of the living stromatolites needed to follow the story of the ancient ones.]

We pull off Highway 101 and

drive into the Shark Bay World Heritage reserve, its thin peninsulas extending like fingers into the Indian Ocean. A carpet of green shrubs covers the rolling land around us—not a single building or fence line is visible. Notices posted along the road sternly warn us to keep to a few well-defined tourist roads and beaches: this is not a place we can roam at will.

Dr. Lindsay Collins, an expert on the stromatolites at Curtain University in Perth, had advised us to stay away from the tourist areas at Shark Bay.

"Call Peter Kopke—he'll let you camp on his beach in Hamelin Pool. Peter's got the best stromatolites in the whole area—in fact, David Attenborough brought his crew there and filmed them for his *Rise of Life* series."

Carbala Homestead

We tour the main reserve briefly, and then head for Peter Kopke's place, Carbala Homestead. When we turn off the main drive onto Carbala Road, the first thing we see is Peter's windmill in the distance, standing straight as a pine tree on this flat land. Then we glimpse the grove of trees that in Australia always marks where we'll find the main house.

Peter Kopke is waiting for us as we drive up to his ranch yard. The late afternoon sun deepens the lines in his face until it resembles the parched land we've been traveling. When he shakes our hands, his calloused palm and fingers feel rough as tree bark. He draws a quick map in the dirt to show us how to get to the beach but refuses any pay for letting us camp on his land.

We navigate the red dirt road that leads us toward Hamelin Pool. Near the beach we find a circle of thick bushes where we park the camper. The wind is blowing off the ocean, and it's good to feel the moisture on our faces and smell the salt air after days in the Outback.

A Landscape Like No Other

At the top of a rise, we catch our first sight of Hamelin Pool and its broad, pristine beaches spreading along the bay. Near the shoreline is a layer of what look like flat plates of stone half submerged in the water. Beyond that, we see the stromatolites themselves—at first a few solitary

columns close to shore, then farther out, larger clusters that merge by the thousands into a stony carpet stretching the length of the bay.

We shoulder our gear and start hiking toward shore, but an odd crunching noise under our boots stops us. Glancing down, we discover we're walking on a layer of white shells, each the size of a fingertip. We look around and yell in shock. The entire beach—several feet deep and spreading for miles on either side of us—is completely made of shells.

We reach the water and see more shells mixed in the silt, slowly turning into limestone. The stone blends with stromatolites colored by red algae. We have stumbled onto an artist's dreamscape.

HAMELIN POOL
AT SHARK BAY

At Mt. Narryer we learned to take our time and let a place sink into us before starting our work. There's a lot we don't know about this site, such as when high and low tides occur. We didn't think to ask Peter. Nor do we know whether we'll have an onshore or offshore wind tomorrow or whether it will be cloudy or clear.

Instead, we simply listen. Shallow waves splash, one of the oldest sounds on Earth. A few gulls cry then fall silent. The wind hisses through salt grass on the hill behind us. Finally we hear the voice we're waiting for tell us where to begin. We pick up our cameras and start working among the stromatolites close to shore. There are only two or three hours of daylight left.

When evening comes, we build a fire on the beach and watch as pearlescent colors spread across the sky. They linger over Hamelin Pool long after the sun has set.

" . . . a dormant power awakening "

The next morning, a brisk wind

and a low tide reveal all the stromatolites submerged the night before. We're charmed by some of their playful shapes; it's easy to believe their tiny builders possess a vigorous artistic vision.

Resting for a while on the beach, we find ourselves talking more openly about our lives. The year before, Sue had helped her mother die peacefully at home, honoring her mother's last request. A few years earlier, Lynn had taken steps to end her 18-year marriage and begin a new life.

We recall old fears that stripped us of our artistic lives for so long. During that time, calling ourselves "artists" always seemed fraudulent, and we had slowly stopped creating our own work. Now, surrounded by these primordial creators, we feel a dormant power awakening in response.

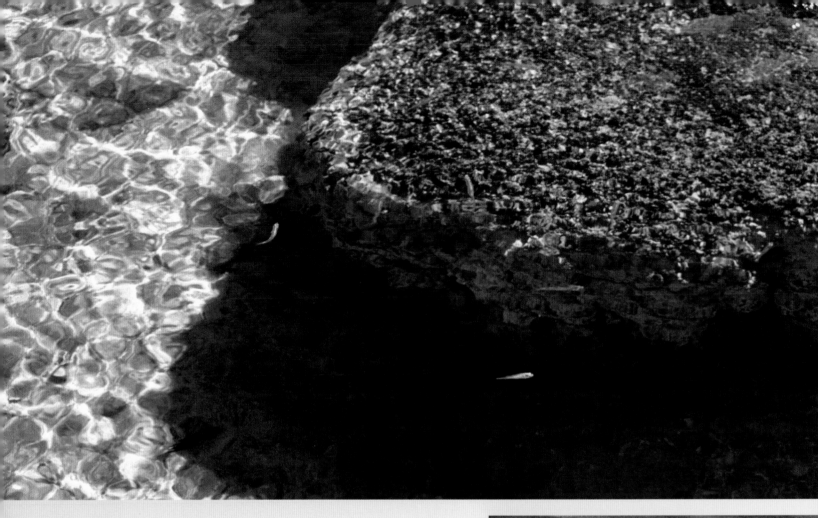

We split up and follow whatever calls us, whether it's the profile of a solitary stromatolite or the jellyfish that pulse nervously as our shadows brush across them. These small creatures, accompanied by fish the size of minnows, appear and disappear like smoke among the stromatolites.

The water is too cold for wading, so we stick a waterproof camera under the surface and snap a few pictures. The blurry focus adds a surreal touch to an underwater stromatolite speckled with light (right).

We feel a growing sense of affection for these bacteria and their squat columns and mats. Even though there's a huge evolutionary gap between us, somehow they evoke a sense of kinship.

We see more stromatolites farther out in the bay and step carefully across the columns to reach them. Absorbed in our work, we forget about the tide.

*T*he stromatolites around us are disappearing underwater. The morning tide is coming in, and it's coming fast. We jump from one dry spot to another as we head back to shore, taking quick photographs along the way. By the time we've shot the last frame of our last roll of film, the entire colony is submerged.

We pack up our film and wonder whether our photo shoot has been successful. Only when we're back home looking at all the developed images will we see how much of our artistic sense we recovered here. It will give us the courage to continue the journey long after our travels are completed.

But for now, gazing at the empty stretch of water covering the stromatolites, we feel an unaccountable sense of loss. For a few hours we bridged a gap not only of species but of time, linked by a common chain of life that after three billion years echoes within us.

We touch the water where the colonies lie hidden, feeling a connection between their filmy layer and our fragile skin. The blue water and blue of the sky enclose everything that rests between; the moment holds still. This is reverence.

When you touch them, all things give you their story.
—David Mowaljarlai, Australian Aboriginal elder

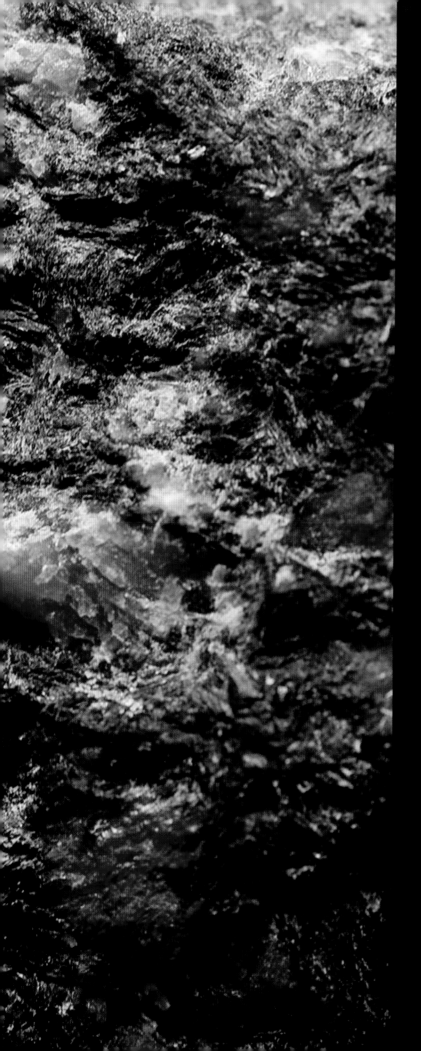

ANCIENT MINERALS
WITHIN US

The strength that lies within our bones ...

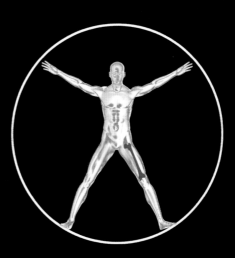

ANCIENT MINERALS
WITHIN US

Gemlike crystals of hydroxy-apatite, made up of calcium, phosphorus, and hydroxide. This is the most abundant form of apatite in human bones.

Previous page: Close-up of Acasta gneiss, showing the stone's structure, which includes the mineral apatite.

As we reflect on our journeys, we discover that the connection we sensed between ourselves and ancient stone, that "feeling in the bones," was more literal than we knew. The mineral apatite, found in the oldest rocks such as Acasta gneiss, makes up most of our compact bones and teeth.

Without apatite, human bones would bend like rubber, unable to bear our weight. Layers of our bone are built and sustained by structures such as the Haversian canals shown below and at right. The canals contain new bone cells, while tiny openings allow blood vessels and nerves to pass through.

Seen up close, these structures echo the wind-carved swirls and hollows of weathered stone. The similarities are fitting. Four and a half billion years ago, minerals were used to build the first rocky crust of Earth and—eons later—to build the solid structures of evolving life.

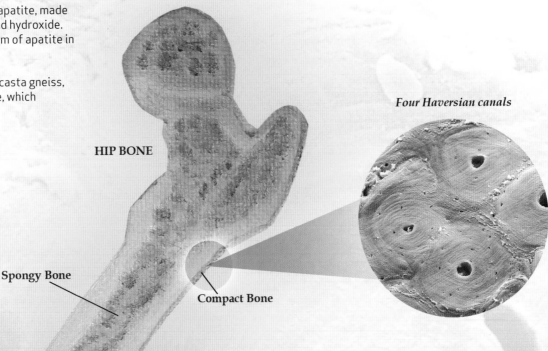

Four Haversian canals

HIP BONE

Spongy Bone

Compact Bone

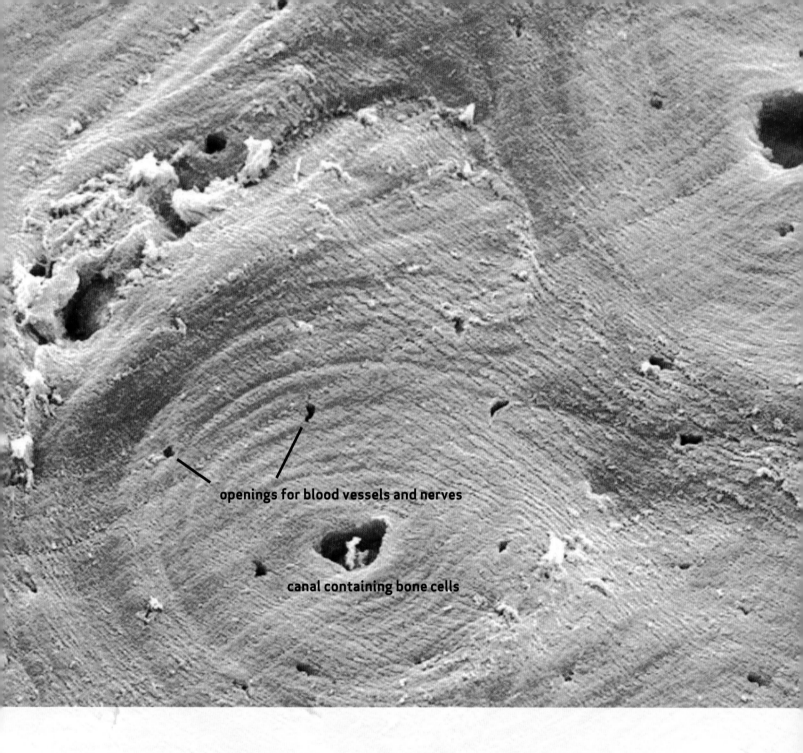

openings for blood vessels and nerves

canal containing bone cells

The ancient minerals are

clues to our beginning;

we carry Earth's history deep within us.

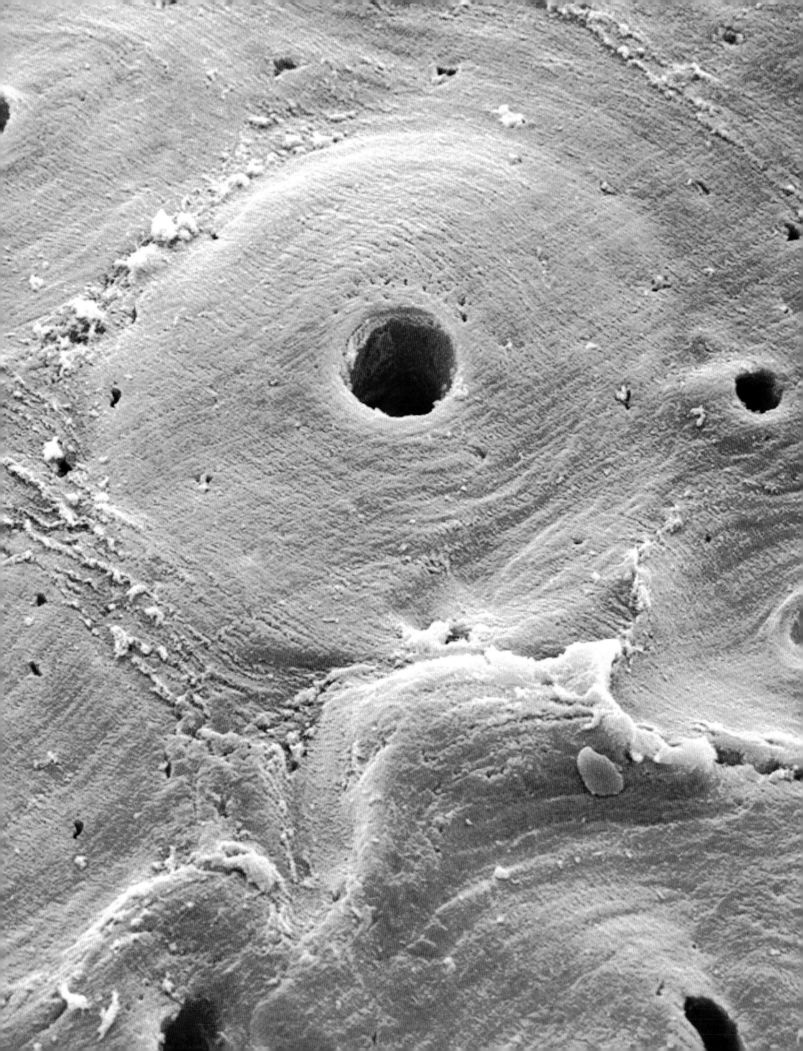

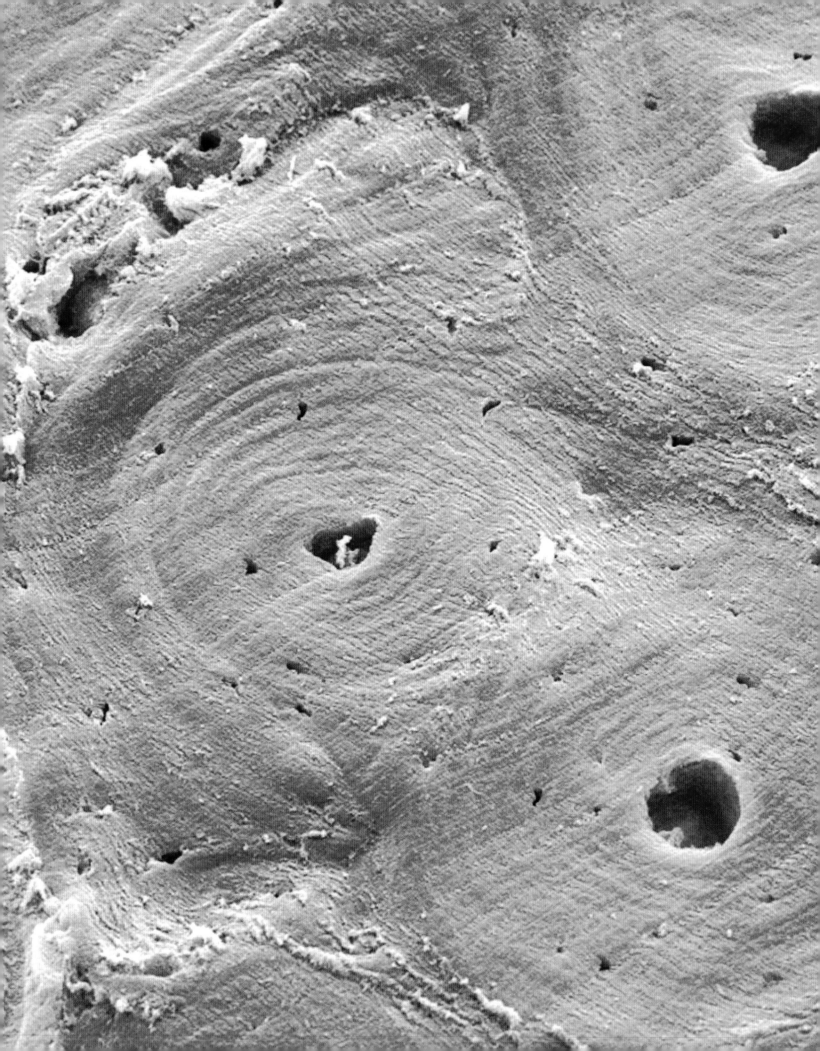

THE STRENGTH WITHIN OUR BONES

Apatite, one of Earth's oldest minerals, was first formed in the chemical alchemy of the young planet 4.5 billion years ago. Our bodies, imitating this ancient process, perform a similar alchemy, combining calcium, phosphorous, and hydroxide to form the apatite used in our skeleton.

To build living bone, nature took layers of apatite with their ability to bend under pressure and to bear immense loads, and then threaded the layers with nerves and blood vessels. It put in a binding filler and gave the layers the ability to repair themselves. Millions of years ago, this living bone allowed life to leave the sea and bear its own weight on land.

We repeat this story in our first year of life. Our bones are softer while we float in the womb for the first nine months; but once we enter the world, our bodies start building the harder, bony skeleton we'll need to live on land.

Mineral Legacy of Bone

In compact bone, apatite crystals are contained in cross layers that allow the bone to bend but not too much. In the inner spongy bone, needle-like crystals form part of a matrix that enables bone to resist compression under stress. Without this flexibility and strength, we would not be able to stand, walk, or move in countless ways.

How Our Bones Endure

But the similarities between stone and our human skeleton go deeper. Just as stone is transformed by heat and pressure into a harder, more resistant structure, so our bones gain thickness and strength from being stressed by weight and movement. As the bone is challenged, more apatite is added to reinforce the layers.

And like rock, bone appears to weather and erode. Even under normal stress, tiny cracks and fissures appear in the bone's outer surface, the same way that stone cracks under heat, cold, and pressure. Bone "erodes" as older layers are reabsorbed by the body or as they thin with age. Unlike stone, however, our bones can repair themselves, using apatite to seal cracks and build new layers.

This ancient mineral is a persistent echo of early Earth within us, one we felt deeply on our travels. But these minerals are not the only connection— ancient life resonates within our bodies as well.

Apatite in Human Bones

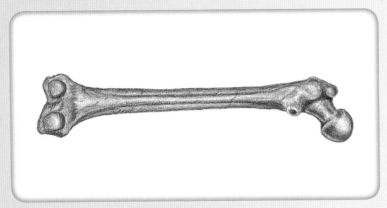

Hip bone with apatite reinforcing its compact and spongy layers is strong enough to support our weight.

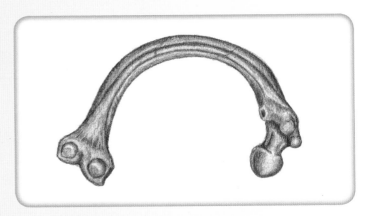

Hip bone without apatite would bend like rubber under stress.

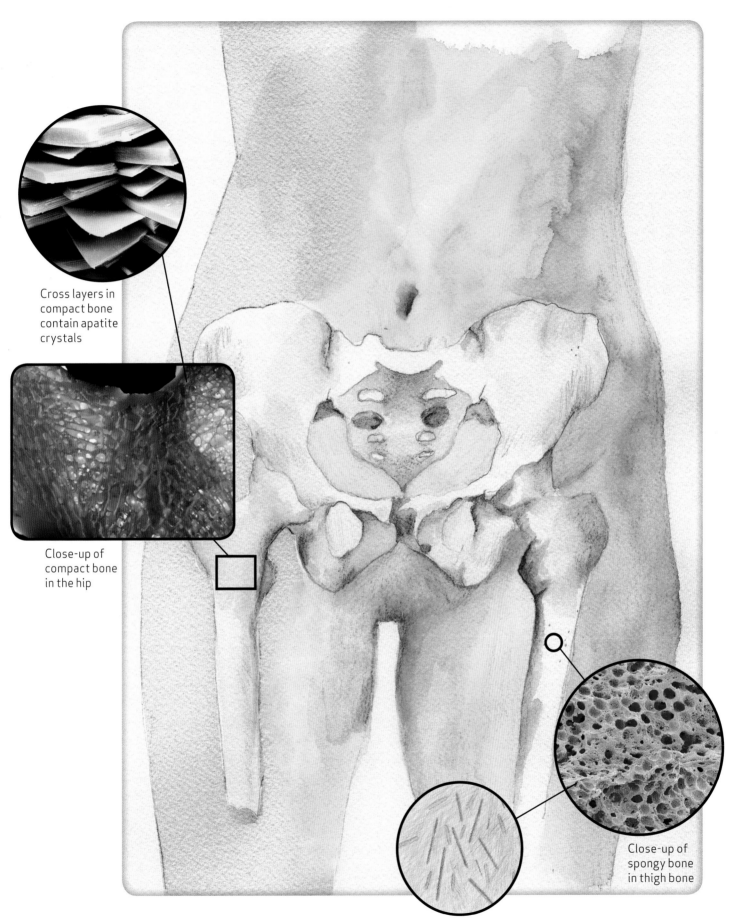

Cross layers in
compact bone
contain apatite
crystals

Close-up of
compact bone
in the hip

Apatite needles in spongy bone

Close-up of
spongy bone
in thigh bone

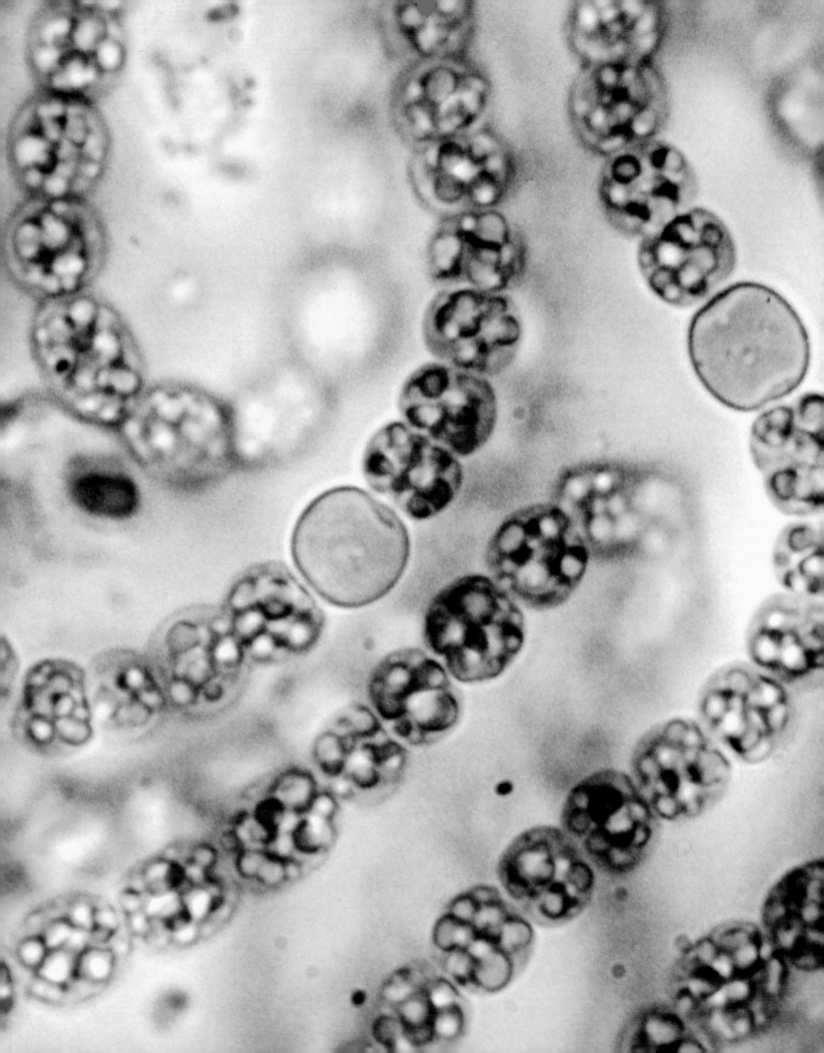

ANCIENT LIFE
WITHIN US

The bonds that keep us alive . . .

ANCIENT LIFE
WITHIN US

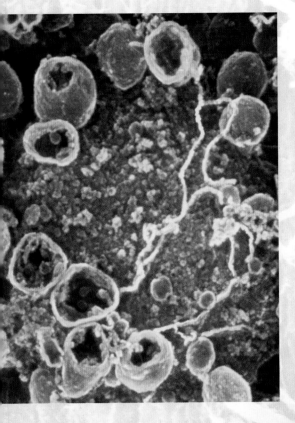

*W*hen our journey began, we did not expect to find so intimate a kinship between humanity and early life. But just as ancient bacteria were vital to shaping the young planet, so many of their descendants are vital to our own survival.

In the close-up of healthy human heart muscle below and at right, the small greenish globes are mitochondria, tiny organelles that convert oxygen and nutrients into energy the body can use. Without them, our hearts would stop beating; our bodies would die.

Yet mitochondria were once free-living bacteria. Nearly 2.5 billion years ago, these bacteria found a way to use oxygen instead of sunlight as fuel. In time, the bacteria were engulfed by larger cells, forming a new type of cell that became the building block for all complex life on Earth, including us. But our true relationship with these tiny lives is much deeper and more complex than we imagine.

Magnified image of a human cell shows mitochondria (green) surrounding the nucleus (gray background). The more energy a cell requires, the more mitochondria it needs to produce that energy.

Previous page: *Nostoc cyanobacteria*, greatly magnified, a modern species that is remarkably similar to ancient cyanobacteria.

Heart muscle and mitochondria

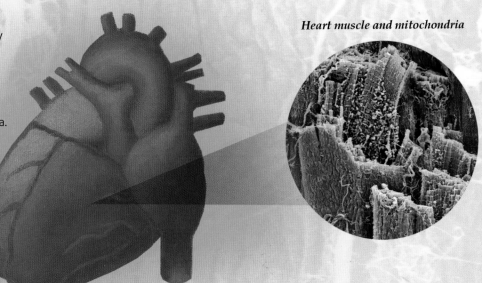

HEART

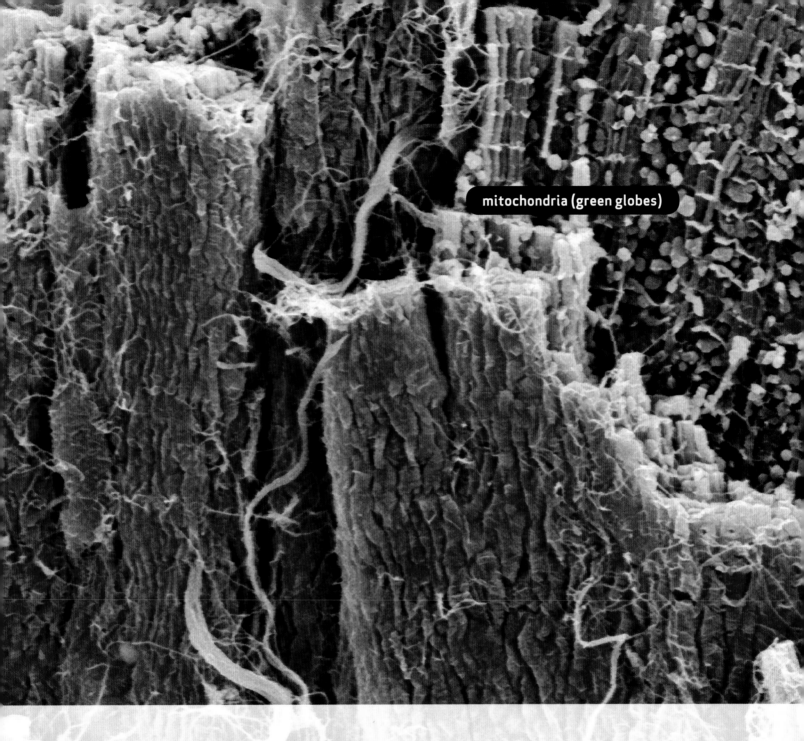

mitochondria (green globes)

Mitochondria are clues to the

ancient bonds that sustain us;

we carry the evolving story of life
within our very cells.

THE BONDS THAT HELP
KEEP US ALIVE

The story of the connection between ancient bacteria and mitochondria began with an odd discovery: Our mitochondria carry their own genetic material—including their own DNA—that is different from the genetic material of other structures in the cell. Why was this the case?

The Puzzle Solved

Several biologists and microbiologists became intrigued by the question. They spent many years researching the genetic history of the mitochondria and found evidence linking them to a species of oxygen-breathing purple bacteria.

At some point, larger cells captured these bacteria, which led to a mutually beneficial arrangement: Cells supplied the bacteria with nutrients and oxygen, and the bacteria converted the materials into energy for their hosts. Over time, the bacteria evolved into mitochondria, the powerhouses that produce the energy in our tissues and cells.

As animals and bacteria evolved over millions of years, they formed other beneficial relationships that our modern human bodies rely on as well.

Bacteria and Human Life

Recent discoveries about the bonds between humans and bacteria reveal an astonishing fact: our bodies are made up of 6 to 10 trillion cells but contain 60 to 100 trillion bacteria and Archea, the earliest forms of life on Earth. Only a few of these species cause disease. Contrary to what we believe, the great majority of them help keep us alive.

Bacteroides fragilis, for instance, break down tough plant fibers and help us absorb vital nutrients. *Lactobacillus salivarius* make two vitamins our body can't produce on its own: a type of vitamin B, important for nerve function, and vitamin K, essential for blood clotting. Still other types of bacteria keep harmful organisms from colonizing our skin. Without these and other species, we couldn't digest our own food, survive even minor cuts, or protect ourselves from the bacteria and viruses that cause illness.

The affection we felt for the cyanobacteria at Hamelin Pool was based in fact, although we didn't know it at the time. Something inside us recognized our link to the oldest life on Earth.

How Bacteria Became Mitochondria

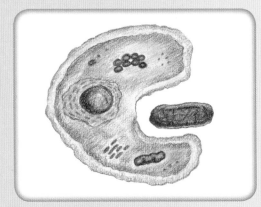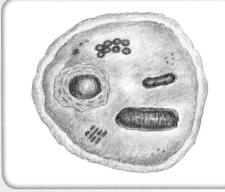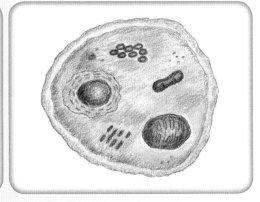

Scientists believe that a species of ancient purple bacteria, which used oxygen for fuel, were engulfed by larger, more complex cells.

Once inside a cell, the bacteria continued to produce energy by using oxygen and nutrients supplied by the cell.

Bacteria gradually evolved into mitochondria, the cell's energy producers, and lost their ability to survive outside the cell membrane.

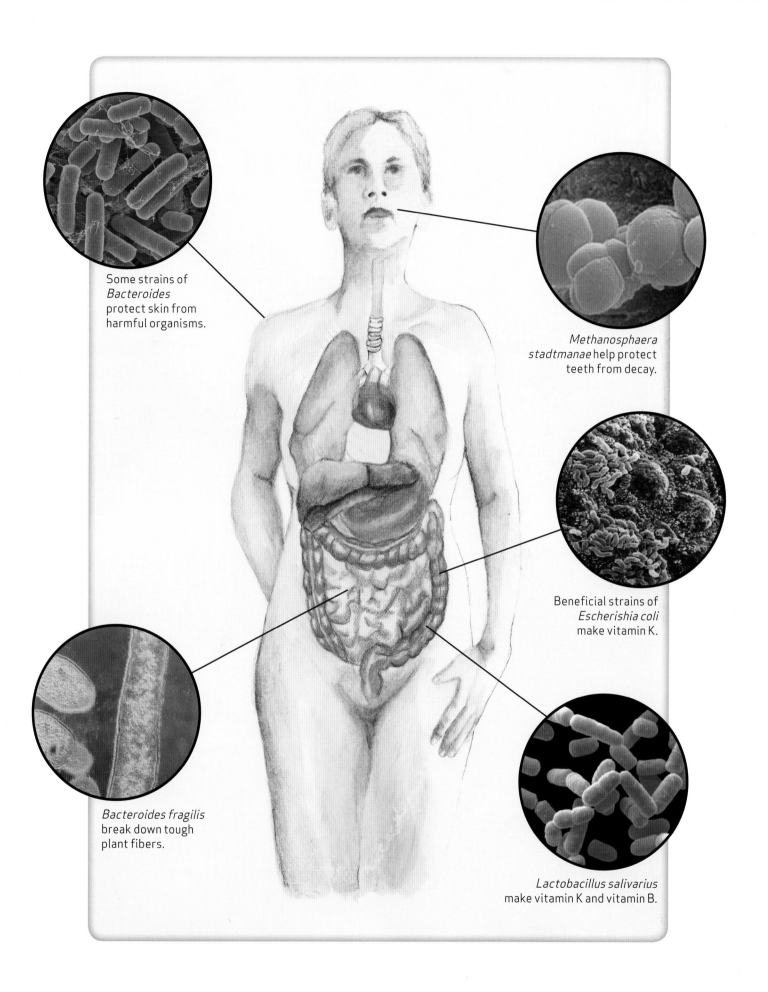

Some strains of *Bacteroides* protect skin from harmful organisms.

Methanosphaera stadtmanae help protect teeth from decay.

Beneficial strains of *Escherishia coli* make vitamin K.

Bacteroides fragilis break down tough plant fibers.

Lactobacillus salivarius make vitamin K and vitamin B.

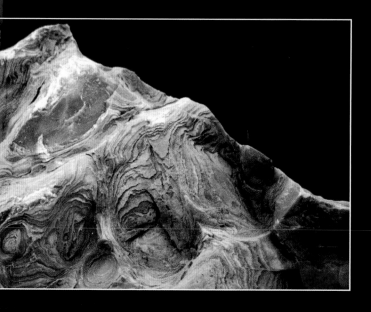

ART & SCIENCE
OF EARTH HISTORY

The story begins in the stones . . .

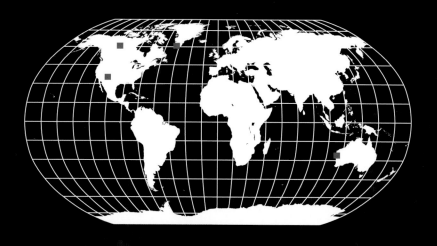

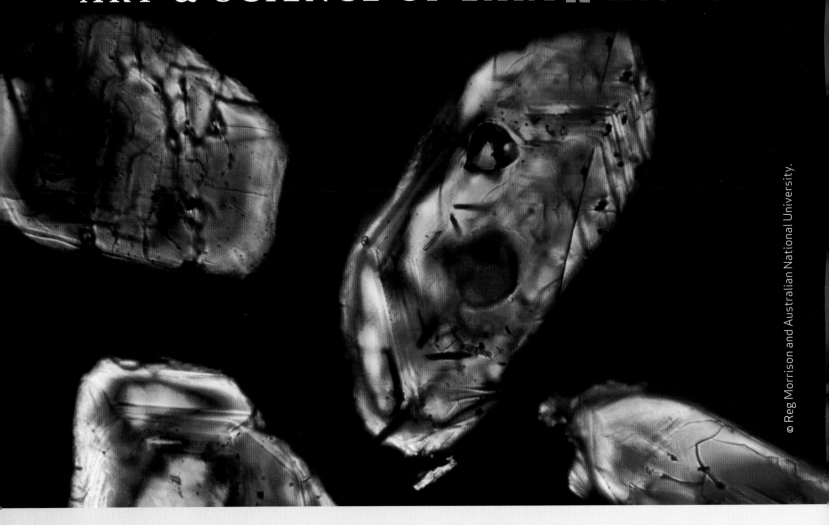

© Reg Morrison and Australian National University.

The story of Earth told by science often conveys as much beauty and mystery as the story we encountered on our personal journeys. The stones speak through the language of science as well as art, enhancing our sense of wonder about the knowledge they convey. In this final chapter, we offer some of the stories science has gleaned from the planet's oldest minerals, oldest skin, and oldest colonies of life. The more we learn about Earth, the more mysterious this living world appears.

EARTH'S OLDEST MINERALS

One of the questions we asked Dr. David Nelson was, "How did you know where to look for the oldest zircons?" He explained: "These crystals are commonly found in ancient stone, and since we knew where the oldest rock layers were, it followed they would have the oldest zircons. In the 1980s and '90s, we hauled hundreds of pounds of this rock from Jack Hills and Mt. Narryer back to the university labs."

Earth's Time Capsules

The hardy zircons are regarded as the gold standard for dating because they can survive relatively intact for eons. As these crystals formed, they sealed trace elements and gases from the surrounding environment within their tough, durable layers. When their original host rock eroded away, the crystals simply waited until younger rocks formed around them. These qualities made zircons the perfect time capsules for early Earth.

But opening the tiny capsules—each one smaller than a grain of beach sand—turned out to be an arduous task. Geologists first had to crush all the ancient rock and then painstakingly sift through the debris to find the best zircon crystals to study.

"Once we cleaned and prepared the crystals," Dr. Nelson said, "we used special ion microprobes to blast a few atoms off their layers." Days of intense, around-the-clock analysis of the vaporized gas and trace elements followed. The final results stunned geologists around the world.

Zircons' Ancient Secrets

The first shock was the sheer age of the Mt. Narryer and Jack Hills zircons—many were over four billion years old. According to the traditional view of Earth's history, no minerals or original crust could have survived the searing temperatures and meteor bombardments of the first 500 million years.

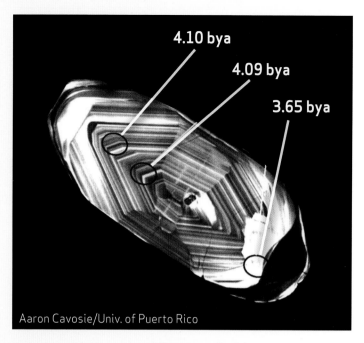

Aaron Cavosie/Univ. of Puerto Rico

A zircon crystal grows by adding layers. Each layer traps gases and trace elements, revealing what conditions were like on Earth when the layer first formed. *bya = billions of years ago*

Yet the evidence was undeniable. These crystals contained radioactive uranium that had decayed at a steady rate over eons, slowly turning into lead. This gave geologists a highly accurate way to measure the zircons' age—the higher the ratio of lead to uranium, the older the zircon. One zircon dated by geologists Dr. John Valley and Dr. William Peck turned out to be 4.4 billion years old.

The second shock was that zircons contained trace elements and a type of oxygen that could be formed only in cooler temperatures and in the presence of seawater. The crystals were saying that a mere 100 million years after Earth formed, it had already cooled enough to acquire crust, shallow seas, and an atmosphere. Cooler, wetter conditions meant that life might have established a foothold on Earth far earlier than anyone believed possible.

Throughout the 1980s and '90s, geologists in Europe and the United States also tested Australia's zircons and confirmed the results. The tiny time capsules had waited more than four billion years to tell a dramatically new story of Earth.

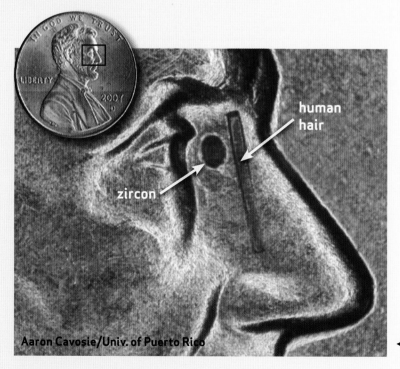

Aaron Cavosie/Univ. of Puerto Rico

◄ This photo shows the true size of most zircon grains. (The box on the penny indicates the area enlarged.)

▲ *What we thought we knew . . .*

Geologists once assumed that for its first 500 million years, Earth was covered by an ocean of fiery magma kept molten and roiling by meteor bombardments. No crust or life forms were thought to survive this hellish time, aptly named the Hadean (or "hell") Eon.

▼ *What we know now . . .*

The zircons tell us that within its first 100 million years, Earth had cooled enough to acquire a solid crust, shallow seas, and an atmosphere. Life may have gained an early start here, perhaps after the meteor bombardments of four billion years ago.

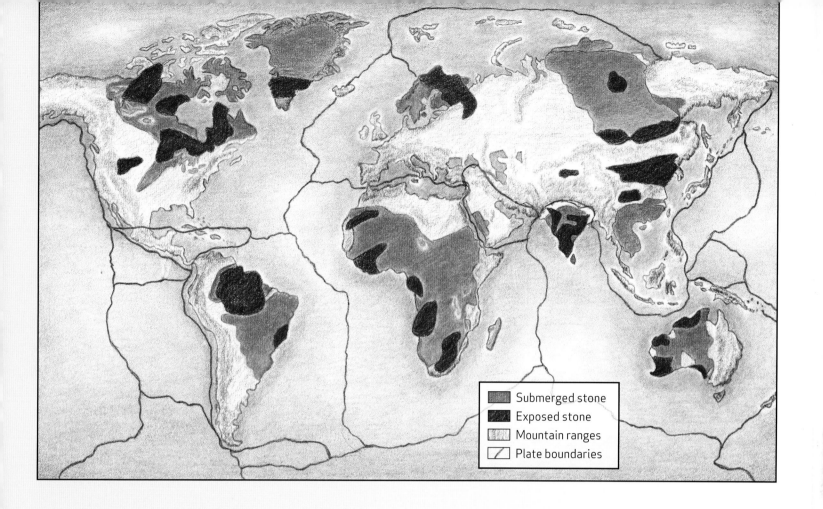

Legend:
- Submerged stone
- Exposed stone
- Mountain ranges
- Plate boundaries

EARTH'S OLDEST SKIN

The map above shows where the oldest skin of Earth lies submerged (orange) or exposed (brown) at the surface, creating a network of ancient stone around the planet. Much of this network was mapped only in the past few decades as geologists searched the world for energy and mineral resources. To their surprise, the heart of nearly every continent contains stone that ranges from two billion to four billion years old.

Since then, geologists have been trying to understand not only how this ancient skin formed but how it has managed to survive the eons. The stone—twice as thick and heavy as younger crust—should have sunk back into the mantle below, yet it remains anchored in the continents. This oldest layer has also been battered by immense forces originating in the planet's fiery interior.

Below the thin layer of crust, Earth's mantle circulates like thick, slowly boiling liquid, kept hot by the super-heated core. Geologists think that currents in the mantle gradually split the crust into massive plates and began shifting and moving them around the globe.

Throughout time, these plates have continued to collide, separate, and grind under, over, and past each other. The oldest skin, caught in their crushing movements, was slowly changed into the mesmerizing colors and formations we found in Australia, Greenland, and North America. These sites are only beginning to yield some of their secrets to geologists.

Earth's Layers (not to scale)

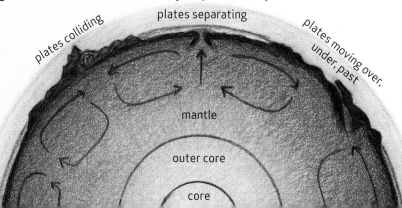

Movements of the Continents Over Time

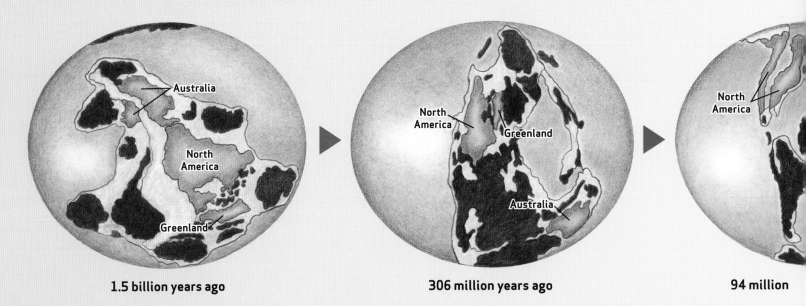

1.5 billion years ago ▶ 306 million years ago ▶ 94 million

Mt. Narryer, Western Australia

For all its travels around the globe, Western Australia remains one of the most stable continental fragments on Earth. Its volcanoes fell silent a billion years ago, and it still retains its ancient blocks of stone, one of which lies under Mt. Narryer and Jack Hills to the north. Yet when geologists dated some of the zircons from these two sites—from 3.5 to 4.4 billion years old—they discovered the crystals were still millions of years older than the rocks that carried them.

This discovery sparked a search for the original bedrock in which the crystals had formed. After nearly three decades, the bedrock has yet to be found. But its fabled existence continues to inspire geologists. As one said, "This is the Holy Grail of geology—a remnant of Earth's earliest crust. To find a piece of it would be like looking straight back to the origins of the planet."

Acasta River, Northwest Territories, Canada

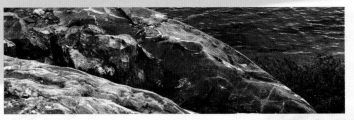

The conditions that formed the four-billion-year-old Acasta gneiss no longer exist on Earth. At that time, the young planet was a hotter, more chaotic place. Geologists know that gneiss started as molten rock expelled from the mantle, but they disagree about how this rock was transformed so soon afterward into the densely layered gneiss.

Because the early core and mantle were hotter, some geologists speculate that mantle currents would have circulated faster, causing plates to collide more violently. The stone, enduring greater cycles of immense heat and pressure, would have been changed more quickly into gneiss.

But how could the stone have survived four billion years of plate movements, severe climate change, and erosion? As yet, no one knows. The answer would help geologists understand how Earth built the surface that all life inhabits.

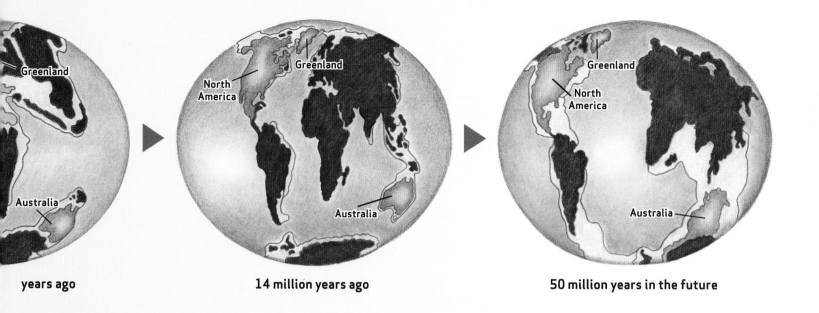

Greenland
Australia
years ago

North America
Greenland
Australia
14 million years ago

Greenland
North America
Australia
50 million years in the future

Akilia Island and Isua, Greenland

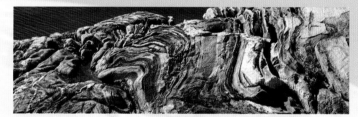

As Greenland was slowly carried from the South Pole, across the equator, toward the North Pole, islands like Akilia broke away from the mainland. Under the massive movements of Earth's plates, Akilia's rock layers endured thousands of pounds of pressure and temperatures up to 3,000 degrees Fahrenheit, which so deformed the 3.8-billion-year-old stone that geologists could barely tell what its original structure had been.

But on the mainland at Isua, northeast of Nuuk, a section of equally old rock has remained largely intact. Even more astonishing, its layers have preserved 3.8-billion-year-old traces of bacterial life, the oldest ever discovered. This meant that life may have appeared either during or soon after the last meteor bombardment in the Hadean Eon. As geologists search for more such traces, the stone seems to be telling us that life is far older and far tougher than anyone ever thought.

Blacktail Canyon and Grand Canyon, U.S.

Blacktail Canyon's Vishnu schist and Great Unconformity are not the only mysteries in the Grand Canyon. Downriver, the 1.7-billion-year-old schist lies above much younger rock layers, while adjoining layers are tilted at odd angles.

Geologists discovered that several million years ago, the entire plateau where the Grand Canyon is located began rising slowly even as the surrounding land crumpled under the pressure of plate movements. The uplift caused earthquakes that rearranged and tilted the rock layers.

The mystery is why the plateau lifted up and still continues to rise from the land around it. Several theories have been proposed, such as bubbles of magma that have been melting rock from the plateau's underside, but none solve the puzzle. One geologist remarked, "It reminds us how much we have to learn about the layers beneath our feet."

EARTH'S OLDEST COLONIES OF LIFE

In a region of Western Australia so remote it was dubbed North Pole, geologists discovered giant fossil stromatolites 3.5 billion years old, the earliest evidence of colony life on Earth. These fossils (top right) are remnants of huge stromatolite reefs, built by bacteria, that once stretched along Western Australia's coast.

When Bacteria Ruled the Earth

For nearly three quarters of Earth's history, bacterial forms of life dominated the planet, constructing their massive colonies around nearly every seacoast and waterway in the world. They lived in groups of several species, slowly building layer after layer of each stromatolite.

The more researchers learned about these builders, the more they realized that bacteria are among the most important organisms Earth has ever known. Species such as cyanobacteria some-how "invented" photosynthesis, using sunlight for energy and producing a vital byproduct, oxygen.

For millions of years, bacterial colonies pumped oxygen into the oceans. As the gas combined with iron in the seawater, the oceans began to "rust," depositing thick layers of red iron ore on the ocean floors. Plate movements uplifted these layers, forming huge iron deposits around the world.

When the oceans became saturated with oxygen, the gas began to escape into the atmosphere and reacted with sunlight to form the ozone layer. Life could now live on land, protected from the sun's lethal ultraviolet radiation. In a span of two billion years, these microscopic organisms helped prepare the planet for the evolution of complex, oxygen-breathing life.

A Legacy Continues

Then roughly 800 million years ago, the great stromatolite colonies began to decline, victims of Earth's cooling climate and a growing number of predators that grazed on the bacteria. Today, the last living colonies are found in only a few places in the world, including Shark Bay; Baja, California; and the Bahamas.

But the ancient bacteria are far from gone. Their descendants are vital to every ecosystem on Earth, including the systems within our bodies. Some species became mitochondria, supplying energy to most animal cells. Others became chloroplasts, which carry on photosynthesis in most plants. Still others break down waste matter and pollutants. Nearly four billion years after they first appeared on Earth, bacteria still maintain a world their ancestors shaped so profoundly.

How Bacteria Build Stromatolites

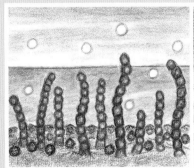

During the day, bacteria carry on photosynthesis, releasing oxygen (white bubbles) and using grains of silt (brown) to build a layer.

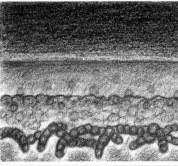

At night, photosynthesis stops, and bacteria become inactive. A layer of silt accumulates until it covers the bacterial colony.

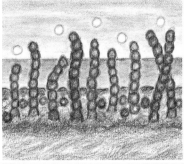

At daybreak, bacteria move up through the silt to reach the sunlight and begin to build a new layer, leaving the old one behind.

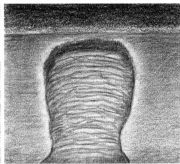

Stromatolites grow only to the water's surface. If the top layer were exposed too long to the sun and wind, the bacteria would die.

© Reg Morrison

Left In the North Pole region, which lies several hundred miles northwest of Mt. Narryer, the Nullagine River exposed a layer of fossils whose stromatolite domes measure nearly six feet in diameter. In contrast, the domes in the much younger Great Slave Lake fossils are only a few inches across.

Below Like their ancestors, modern bacterial colonies build a variety of stromatolite shapes, such as these found at Hamelin Pool in Shark Bay. They range from single columns to joined clusters to a continuous mat. The Hamelin Pool stromatolites range from one thousand to three thousand years old. ▼

INTRODUCTION

The timelines on pages 178-183 offer another way to grasp the immense time frame of the planet's history and to understand the story of the stones.

The **Typical Timeline** (far right), commonly found in books and on web pages, minimizes the first four billion years of Earth's history. This timeline focuses only on the rise of complex life and the appearance of humans.

In contrast, the **Full Timeline** (below) shows that in the first four billion years, ancient stone and ancient life transformed Earth's environment and enabled complex life to arise. This story extends through the Hadean, Archeon, and Proterozoic eons.

HADEAN EON (4.6 to 4.0 bya)

- Sun, solar system, planets, and moons form

- Earth cools, oldest zircons form in crust; planet acquires shallow seas and atmosphere

ARCHEON EON (4.0 to 2.5 bya)

- Acasta gneiss forms in continental fragments

- Crust breaks into moving plates, carried by currents in Earth's mantle

- Amino acids/DNA, building blocks of life, appear

- ARCHEA organisms (first domain); earliest life on Earth evolve

- EUBACTERIA organisms (second domain), evolve. Some eubacteria use photosynthesis, pumping oxygen into oceans and atmosphere

PROTEROZOIC EON (2.5 bya to 542 mya)

- Other eubacteria begin to use oxygen for fuel, which suggests mitochondria exist

- EUKARYOTE organisms (third domain) evolve— ancestors of all complex life

- Multicellular life, male/female gender, and sexual reproduction all evolve

Full Timeline: 4.6 billion years ago (bya) to present

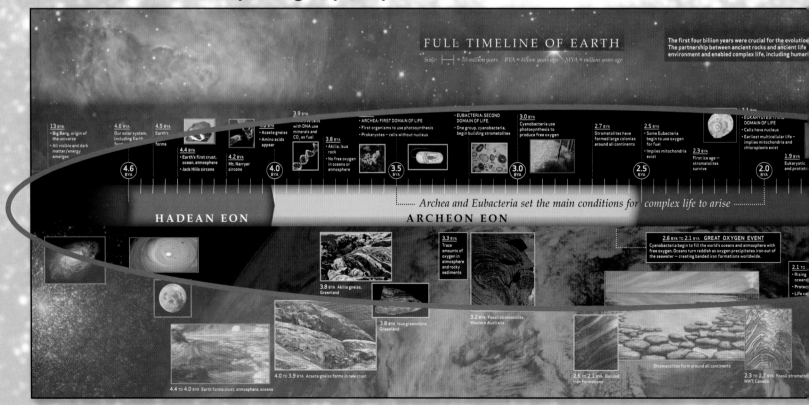

Typical Timeline: 4.6 billion years ago (bya) to present

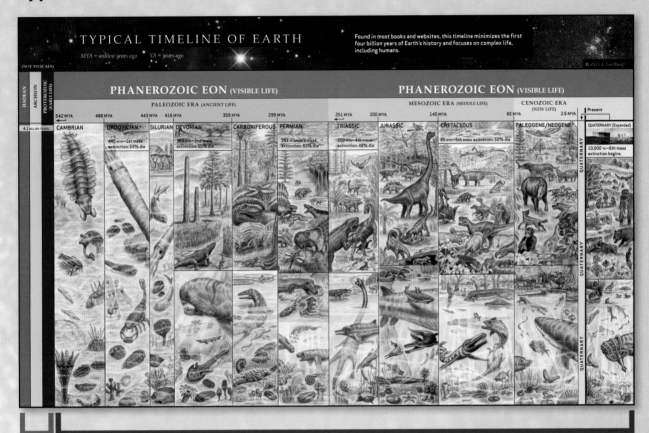

TYPICAL TIMELINE OF EARTH

MYA = million years ago   YA = years ago

Found in most books and websites, this timeline minimizes the first four billion years of Earth's history and focuses on complex life, including humans.

© 2011 L Sue Baugh

(NOT TO SCALE)

HADEAN	ARCHEON	PROTEROZOIC (EARLY LIFE)

PHANEROZOIC EON (VISIBLE LIFE)   PHANEROZOIC EON (VISIBLE LIFE)

PALEOZOIC ERA (ANCIENT LIFE)   **MESOZOIC ERA** (MIDDLE LIFE)   **CENOZOIC ERA** (NEW LIFE)

542 MYA   488 MYA   443 MYA   416 MYA   359 MYA   299 MYA   251 MYA   200 MYA   145 MYA   65 MYA   2.5 MYA   Present

4.1 BILLION YEARS

CAMBRIAN   ORDOVICIAN   SILURIAN   DEVONIAN   CARBONIFEROUS   PERMIAN   TRIASSIC   JURASSIC   CRETACEOUS   PALEOGENE/NEOGENE   QUATERNARY (Expanded)

445 MYA—1st mass extinction: 59% die

365 MYA—2nd mass extinction: 50% die

251 MYA—3rd mass extinction: 83% die

200 MYA—4th mass extinction: 48% die

65 MYA—5th mass extinction: 50% die

10,000 YA—6th mass extinction begins.

QUATERNARY

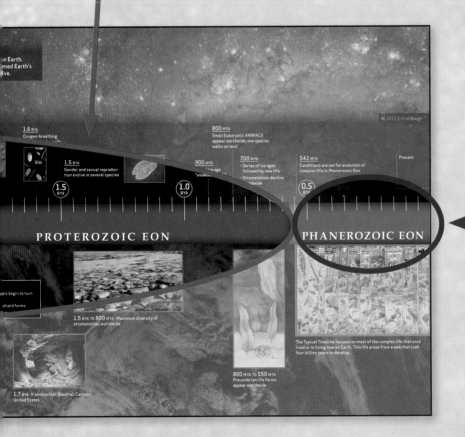

© 2011 L Sue Baugh

1.6 BYA
Oxygen-breathing

1.5 BYA
Gender and sexual reproduction evolve in several species

800 MYA
Small Eukaryotic ANIMALS appear worldwide; one species walks on land

900 MYA
• Series of ice ages followed by new life
• Stromatolites decline ...ldwide

700 MYA
...ice age
...snow...

542 MYA
Conditions are set for evolution of complex life in Phanerozoic Eon

Present

(1.5 BYA)   (1.0 BYA)   (0.5 BYA)

PROTEROZOIC EON   **PHANEROZOIC EON**

...ls begin to turn

...shield forms

1.5 BYA TO 800 MYA Maximum diversity of stromatolites worldwide

The Typical Timeline focuses on most of the complex life that once lived or is living now on Earth. This life arose from a web that took four billion years to develop.

800 MYA TO 550 MYA
Precambrian life forms appear worldwide

1.7 BYA Vishnu schist, Blacktail Canyon, United States

PHANEROZOIC EON (542 mya to present)

- Complex life colonizes Earth's seas, land, and air.

- Human species appear two to three million years ago. Modern humans appear only about 300,000 years ago.

The **Typical Timeline** focuses only on the Phanerozoic Eon, giving the impression that the first three eons of Earth history were not important.

You'll find a different story when you open the **Typical Timeline** on the next pages to reveal the **Full Timeline**. Species such as ours arose from a complex web that took four billion years to develop, a web that will keep producing new species until Earth no longer supports life.

FULL TIMELINE OF EARTH

Scale: ⊢——⊣ *= 50 million years*   *BYA = billion years ago*   *MYA = million years ago*

3.5 BYA
- ARCHEA: FIRST DOMAIN OF LIFE
- First organisms to use photosynthesis
- Prokaryotes – cells without nucleus

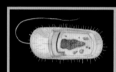

xygen
or
ere

3.2 BYA
- EUBACTERIA: SECOND DOMAIN OF LIFE.
- One group, cyanobacteria, begin building stromatolites

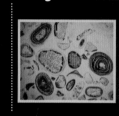

3.0 BYA
Cyanobacteria use photosynthesis to produce free oxygen

2.7 BYA
Stromatolites have formed large colonies around all continents

3.5 BYA   **3.0 BYA**   **2.⬤ BY**

.... *Archea and Eubacteria set the main conditions for*

ARCHEON EON

ia gneiss,

3.3 BYA
Trace amounts of oxygen in atmosphere and rocky sediments

3.8 BYA Isua greenstone, Greenland

3.2 BYA Fossil stromatolite, Western Australia

2.6 TO 2.1 BYA Banded iron formations

n new crust

13 BYA
- Big Bang, origin of the universe
- All visible and dark matter/energy emerges

4.6 BYA
Our solar system, including Earth, forms

4.5 BYA
Earth's Moon forms

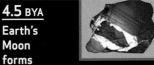

4.4 BYA
- Earth's first crust, ocean, atmosphere
- Jack Hills zircons

4.2 BYA
Mt. Narryer zircons

4.0 BYA
- Acasta gneiss
- Amino acids appear

3.9 BYA
Primitive cells with DNA use minerals and CO_2 as fuel

3.8 BYA
- Akilia, Is[...] rock
- No free [...] in oceans[...] atmosph[...]

(**4.6** BYA)   (**4.0** BYA)

HADEAN EON

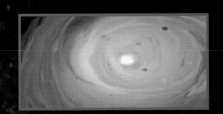

3.8 BYA Aki[...]
Greenland

4.0 TO **3.9** BYA Acasta gneiss forms

TYPICAL TIMELINE OF EARTH

(NOT TO SCALE)

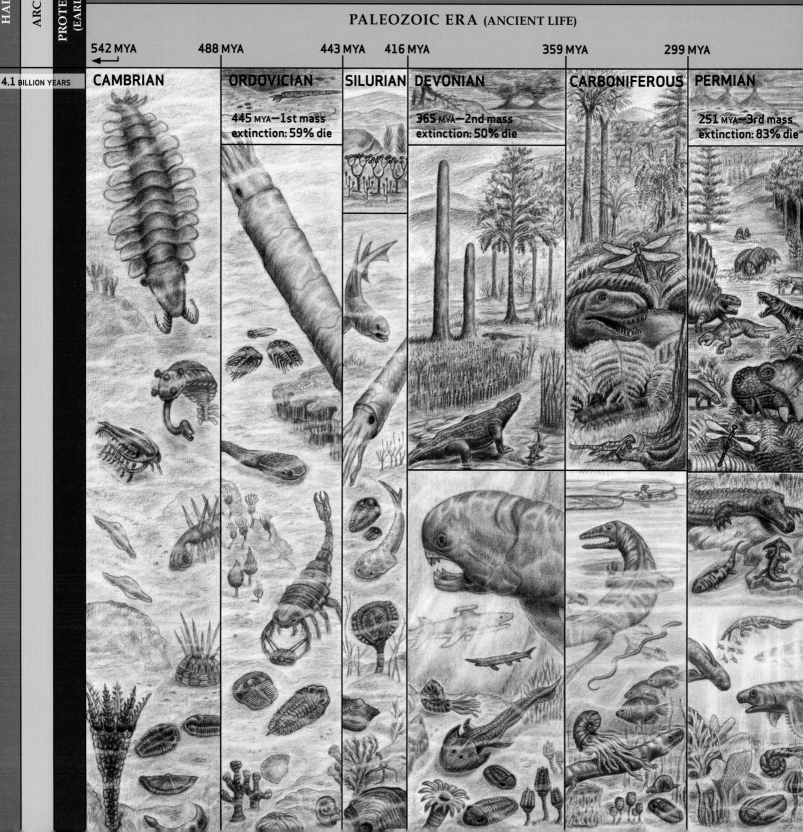

HADEAN

ARCHEON

PROTEROZOIC (EARLY LIFE)

PHANEROZOIC EON (VISIBLE LIFE)

PALEOZOIC ERA (ANCIENT LIFE)

| 542 MYA | 488 MYA | 443 MYA | 416 MYA | 359 MYA | 299 MYA |

4.1 BILLION YEARS

CAMBRIAN

ORDOVICIAN

445 MYA—1st mass extinction: 59% die

SILURIAN

DEVONIAN

365 MYA—2nd mass extinction: 50% die

CARBONIFEROUS

PERMIAN

251 MYA—3rd mass extinction: 83% die

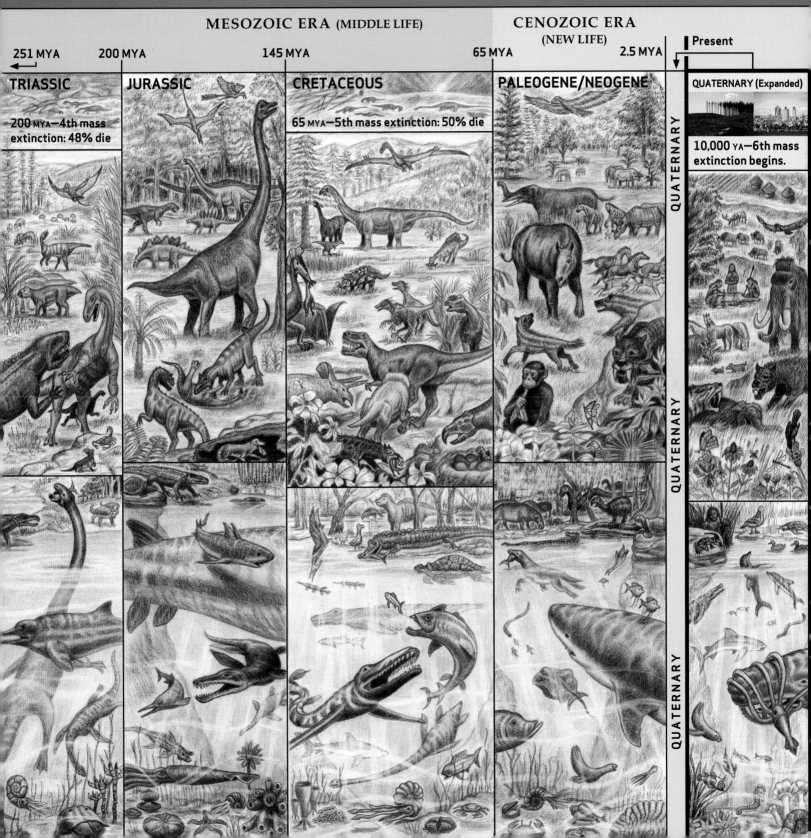

Found in most books and websites, this timeline minimizes the first four billion years of Earth's history and focuses on complex life, including humans.

PHANEROZOIC EON (VISIBLE LIFE)

MESOZOIC ERA (MIDDLE LIFE)

CENOZOIC ERA (NEW LIFE)

Present

251 MYA | 200 MYA | 145 MYA | 65 MYA | 2.5 MYA

TRIASSIC

200 MYA—4th mass extinction: 48% die

JURASSIC

CRETACEOUS

65 MYA—5th mass extinction: 50% die

PALEOGENE/NEOGENE

QUATERNARY (Expanded)

10,000 YA—6th mass extinction begins.

QUATERNARY

QUATERNARY

QUATERNARY

800 MYA
Small Eukaryotic ANIMALS
appear worldwide; one species
walks on land

900 MYA
Global ice age
"snowball Earth"

700 MYA
• Series of ice ages
  followed by new life
• Stromatolites decline
  worldwide

542 MYA
Conditions are set for evolution of
complex life in Phanerozoic Eon

Present

1.0
BYA

0.5
BYA

PHANEROZOIC EON

...num diversity of

800 MYA TO **550** MYA
Precambrian life forms
appear worldwide

The Typical Timeline focuses on most of the complex life that once
lived or is living now on Earth. This life arose from a web that took
four billion years to develop.

The first four billion years were crucial for the evolution of life on Earth. The partnership between ancient rocks and ancient life transformed Earth's environment and enabled complex life, including humans, to evolve.

5 BYA
- ome Eubacteria
- egin to use oxygen
- or fuel
- mplies mitochondria
- xist

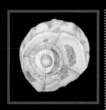

2.3 BYA
First ice age — stromatolites survive

2.1 BYA
- EUKARYOTES: THIRD DOMAIN OF LIFE
- Cells have nucleus
- Earliest multicellular life – implies mitochondria and chloroplasts exist

2.0 BYA

mplex life to arise ....

1.9 BYA
Eukaryotic algae and protists appear

1.6 BYA
Oxygen-breathing Eubacteria diversify

1.5 BYA
Gender and sexual reproduction evolve in several species

1.5 BYA

PROTEROZOIC EON

2.6 BYA TO 2.1 BYA GREAT OXYGEN EVENT
Cyanobacteria begin to fill the world's oceans and atmosphere with free oxygen. Oceans turn reddish as oxygen precipitates iron out of the seawater — creating banded iron formations worldwide.

2.1 TO 1.9 BYA
- Rising oxygen levels begin to turn oceans, sky blue
- Protective ozone shield forms
- Life can now survive on land

1.5 BYA TO 800 MYA Maxi stromatolites worldwide

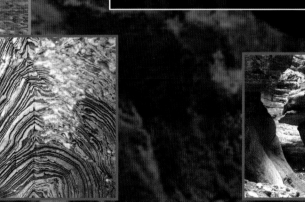

Stromatolites form around all continents

2.3 TO 1.7 BYA Fossil stromatolite, NWT, Canada

1.7 BYA Vishnu schist, Blacktail Canyon, United States

To document the sites, we used single lens reflex 35-mm Minolta, Canon, and Nikon cameras with macro and telephoto lenses; UV and polarizing filters; and Kodak Gold and Fuji color film. With no digital technology, we had to wait until the films were developed to know whether our photo shoots had been successful. This made for some tense moments. The negatives were drum scanned, a technology that extracts the maximum amount of information from the film, to produce the highest quality images for the book.

MT. NARRYER, WESTERN AUSTRALIA

We spent four days at Mt. Narryer, which is located in the Murchinson Shire district of Western Australia. We arrived in the middle of Australia's winter. The mountain is currently protected from mining interests, thanks to the efforts of Sandy and Carol McTaggart.

Foreword, pp. x-ix. We had climbed halfway up the mountain when Sue took this photo of our camper car. It shows the true scale of humans in nature. Perth lies 350 miles to the south, while the coast is 150 miles to the west. The nearest city is about 300 miles to the north, and if you look straight east—you won't see another city for 2,000 miles. For medical emergencies, we would have to use the shortwave radio in our car to call the Flying Doctors, who were 200 miles away by airplane.

Pp. 5-6 We had always wanted an aerial shot of Mt. Narryer and eventually found this photo by Reg Morrison, an Australian photographer. Flying over Mt. Narryer in Sandy McTaggart's plane, Reg used an IMAX camera to capture the mountain's undulating shape, illuminated by the rising sun. We camped close to the northern peak (top of the photo) next to a small red mesa.

Pp. 20-21 After the immense vistas of the Outback, focusing on these rock faces seemed a more intimate way to know the mountain. The stones reveal a great deal about the history of Mt. Narryer and of Western Australia. Granite, feldspar, and basalt (B, D) reflect volcanic activity that ceased about one billion years ago. Quartzite (A, C) was formed under great heat and pressure, while sediments laid down by ancient rivers gradually became sandstone (E).

Pp. 22-23 Lynn is nearly invisible on top of the mesa, blending into the color of the stone. The more time we spent at the site, the more akin we felt to these small mesas, which surround the base of Mt. Narryer. They made a sheltering place to warm up in the morning. At night, even with Carol's blankets wrapped around us, we froze in our camper car as temperatures fell to the low 40s or high 30s.

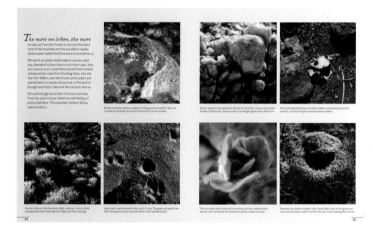

Pp. 24-25 We noticed how Mt. Narryer creates its own environment and even shapes the local weather. Winds shift direction several times a day as the air heats and cools. When the wind sweeps down the mountainside, spiders and magpies catch the insects carried down the steep ravines. In the creek beds and along runoff channels, plants catch and hold the rainwater that flows off the mountain during storms. In turn, the plants support a chain of animals from ants to wild goats. In the four days we were there, we witnessed the rhythms and constrains the mountain imposes on the life around it.

Pp. 28-29 One of the most remarkable moments of our journey was seeing quartz stones glow in the moonlight and mirror the constellations overhead. Sights like this might have been the source of the Aboriginal belief that if you change something on Earth, you change it above as well.

We weren't able to take a photo of the scene, so we combined one of our Mt. Narryer photos with an image of Australia's winter sky taken by New Zealand artist Christopher Picking. When our designer changed our color photo to black and white, the quartz stones glowed white without any retouching, looking just as they had on that moonlit winter night.

AKILIA ISLAND, GREENLAND

Akilia Island is protected as a scientific sanctuary by Greenland, so we weren't allowed to stay overnight on the island. As a result, we had only one day to get all the images we needed. The brilliant sunlight was a true gift for our photo shoot.

Pp. 36-37 This image in particular shows the spirit of Akilia Island. Sue took this photo looking down on the formation from a slight rise. The morning light perfectly illuminated all the curves, colors, and depressions in this breath-taking scene, framed by the blue seawater behind it.

The ripples and curves in the formation reveal where its rock layers softened and folded under the pressure of one plate pushing against another. Rainwater has formed pools in the depressions, and bands of orange and green algae bloom along the moist rock.

Pp. 40-41 Akilia may be a remote location to us, but for people like Ludvig, the guide who ferried us to the island, Akilia is part of their backyard. Scientists share the site with fishermen and seal hunters who use these islands as resting places. The seawater remains so cold year around that many people who sail into open waters don't bother with lifejackets or life preservers. Anyone who falls into the icy water usually doesn't live long enough to be rescued.

Pp. 44-45 This photo captures the beautiful convergence of pattern, light, and color. But the most remarkable fact about this rock pattern is that it was originally formed when the boulder was standing upright. The light bands of quartzite and darker bands of hornblende and feldspar were intruded and compressed like layers of a cake. As the island buckled and heaved under the pressure of plate movements, the boulder was gradually pushed on its side.

Pp. 48-49 Akilia was our first encounter with how artistically nuanced nature can be when transforming stone, from the broad strokes of a dark hornblende dike (top left) to the more delicate lines of quartzite (bottom left). The rippled bands (bottom center) have all the appearance of deliberate artwork, while the stone "foot" (bottom right) adds a touch of humor to the scene. Flame-red lichen draw the eye to a solitary boulder (top right) stranded near the water's edge. Lynn in particular found her subject matter in the smaller, more delicate patterns.

Pp. 56-57 Akilia appeared to be put together from pieces of totally different islands. The top left images show sandstone formations and a "waterfall" of stone that look like spirits carved them. On the bottom left, ancient granite boulders lie half buried in the ground.

The bottom right photo shows a section of Nuuk on the mainland. An island formation serves as a dramatic backdrop to the city. In the old days, Nuuk (meaning "headland" in Inuit) was a place where people gathered to celebrate the summer solstice, trade their goods, and prepare for the long winter ahead.

Pp. 58-59 This aerial shot of Greenland's west coast was the result of a great camera angle and a clean airplane window. You can see melting glaciers at the lower left and top right of the photo. During summer, these rivers of ice deposit sediments (cloudy areas) and calve huge icebergs (white specks) into the fjord. Recently, Greenland's glaciers, part of a massive ice sheet that covers 90 percent of the island, have been melting at a record rate. This is one of the profound changes transforming the Arctic regions.

BLACKTAIL CANYON, UNITED STATES

Our rafting trip down the Colorado River began at Phantom Ranch, the half-way point through the Grand Canyon, and ended six days later at Whitmore Wash. We had about a four-hour stop at Blacktail Canyon. You could easily spend a week in this powerful place.

Pp. 64-65 After a short walk through dry brush and sweltering heat, we entered the narrow passageway of Blacktail. Encountering the dark, powerful Vishnu stone was like stepping into a deep cave. The group, which had been talking and laughing up to that point, slowly fell silent. The presence here reached out viscerally, quieting the mind like a still pool.

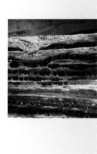

Pp. 72-73 It's easy to imagine some prehistoric animal clawing deep, horizontal grooves in the walls of the canyon. In the smaller photo, the honeycomb pattern indicates this stone has eroded from the inside out. The porous rock allows water to seep in and slowly eat away the softer limestone and sandstone. Over time, the water gradually carves a series of tunnels that eventually reach the surface, creating holes that resemble a honeycomb.

Pp. 76-77 The Vishnu schist, shown here, is aptly named. The Hindu god Vishnu is known as the Preserver, just as this stone preserves nearly two billion years of Earth history. Vishnu is one of the three aspects of divine energy in the Hindu religion, the other two being Brahma the Creator and Shiva the Destroyer. The Vishnu schist is named after the Vishnu Temple formation, which is easily visible from the Grand Canyon rim.

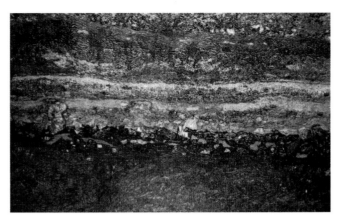

Pp. 78-79 The Great Unconformity is at eye level in Blacktail Canyon, but it is also found outside of Blacktail. High up on the rock walls of the Grand Canyon's Inner Gorge, you can see the Vishnu schist cutting through younger rock layers. Movements in the Earth's crust forced the schist into the sandstone—a reminder that the planet is constantly in motion even if we rarely feel the ground moving beneath us.

ACASTA RIVER, CANADA

The round trip between Yellowknife and the Acasta site was one of the roughest flights we'd ever experienced. Our floatplane was constantly buffeted by thermal winds rising off the tundra. Once we reached Acasta, we had about four hours to work before dark rain clouds closed over the site.

Pp. 83-85 The colors of the greenstone repeated in the rippled bay caught Lynn's eye as she took this photo. The foreground captures some of the glacier's signature effects: deep scratches gouged in the stone and a jumble of smaller rocks and debris left behind as the ice retreated. Unlike Mt. Narryer and Akilia, the stone here was much smoother, feeling almost like dolphin or whale skin. Ironically, repeated ice ages have helped preserve the stone against the ravages of erosion.

Pp. 88-89 The tundra stretches like an ocean of land and water, covering thousands of square miles in Canada. It reminded us of the Outback in Australia. Floatplanes are often the only way remote communities stay connected to the outside world. Much of this vast, uninhabited region is so pristine you can still drink from the lakes and rivers. In summer, evaporation from the tundra creates a cloud cover that can last for days and gives rise to thermal winds that make for rough, stomach-churning airplane flights.

Pp. 92-93 Scientists had to clear the site of willow shrubs, lichen, and moss to study the Acasta gneiss and other rock formations. The cleared area is only about 30 feet by 50 feet but contains a vivid display of ancient stone. The black willow shrubs surrounding the site grow no higher than two to three feet. We found that the farther north we traveled, the more closely the plants hugged the ground to survive the harsh winters.

Pp. 96-97 Viewed through a close-up lens, this formation looks like a primitive landscape in which glacier scratches resemble lightning strikes. The thick, ropey quartzite was formed when the greenstone split under pressure and heat, allowing the lighter mineral to intrude into the crack. Eventually, the rock cooled and "froze" the quartzite into a miniature dike.

Top We were drawn to the almost Zen-like markings on the stones, created by the same immense forces that can melt solid rock. **Bottom** Glaciers forced these layers together, almost as if they were experimenting with contrasts in color and texture.

98

Pp. 98 In the two top photos, lighter minerals look like inlays set in the original stone. Lynn kept finding these small treasures among the larger formations, left like grace notes in the rock.

In the bottom photo, Sue framed the image on the diagonal to emphasize the sharp contrast of red and black stone and the line of pebbles separating them. The red stone probably contains hematite, known for its reddish pigment, while the bottom stone is mostly darker feldspar. Glaciers helped create these formations, often pushing different layers of stone on top of one another to produce such striking contrasts.

Pp. 101–102 We shot this formation from several angles before we finally saw the "waterfall" in the stone. In the book, we used a vertical foldout to reflect this change in perception. At first, only the bottom half of the formation is visible, which is impressive in itself. But when you lift the foldout, the waterfall of stone flows down the page. Time and again, the sites spoke to us in ways that reawakened our own artistic voices.

Pp. 106–107 Lynn spent part of the time photographing the intricate patterns of lichen, mosses, fungi, and other plants surrounding the site. In the brief Arctic summer, plants and fungi have to flower, fruit, and seed quickly. If you study this image closely enough, you can see some of these stages taking place in only a few square inches of ground. The dense tangle of life in this photo is all the more astonishing considering the land is locked in ice and snow for most of the year.

GREAT SLAVE LAKE, CANADA

Rough weather prevented us from taking a trip out to Blanchet Island, but the overcast skies worked in our favor when we got to Captain Smith's backyard in Yellowknife. The dim light helped frame our shots of the fossil.

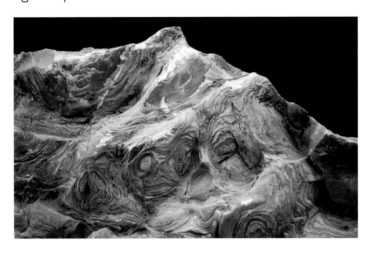

Pp. 114-115 The top of the slab is perfectly outlined against the darker background of the fence bordering Captain Smith's property. During our trip, we were surprised to learn that Yellowknife does not protect these rare fossils. Collectors can simply hire someone to go out to the island, carve out a section or slab, and haul away as much as they like.

Pp. 118-119 We took photos of the lichen from different angles, showing their circular patterns and intricate lines tracing the layers of the fossil built by bacteria.

These stromatolites originally formed when a large, inland sea cut a huge swath through what is now North America. Its warm, shallow waters allowed the bacteria to thrive. Where the water was too murky for sunlight to penetrate, the bacteria lived by using fermentation, a way to make food that doesn't require sunlight.

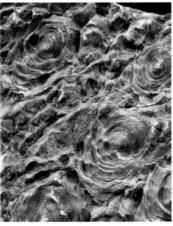

Pp. 120–121 The large photo shows cross sections of the domed structures the bacteria built 1.7 billion years ago. These domes are only a few inches across, compared to those in the 3.5-billion-year-old Australian fossils that measure up to six feet across.

The size differences reflect changes in the global environment. As Earth's climate cooled and more predators arose, the bacteria struggled to maintain their colonies. These conditions helped end the reign of bacteria as the dominant form of life on Earth.

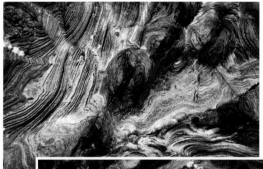

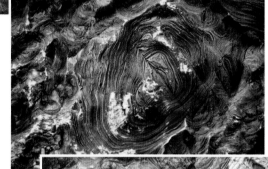

Pp. 122-127 For this series, Sue used a macro lens to focus on smaller and smaller sections of the slab. Yet the close-ups give the opposite impression—the tighter the focus, the larger the structures appear.

These photos turned out to be a rich source of artistic images. Isolating different parts of a photo or changing the orientation yielded multiple portraits within the larger formation.

It's hard to escape the impression that some creative matrix underlies not only these forms but the organisms who built them. That matrix may have been passed down to us, perhaps encoded in our DNA.

SHARK BAY, AUSTRALIA

We spent a day and a half at Mr. Kopke's beach under perfect conditions—full sunlight, a strong offshore wind, and a waning moon. We learned later that these conditions produced unusually low tides. Not only did these tides give us more time to work, but they also exposed more stromatolites for us to photograph.

Pp. 131-132 These stromatolites enjoy a sheltered environment in Hamelin Pool—the water is so salty few predators can survive.

We didn't fully appreciate what we found here until after we photographed the Great Slave Lake fossil stromatolite in Canada. The history of that fossil taught us the value and significance of the living stromatolites. This aspect of the journey caught us by surprise; we had not planned to learn about ancient life, only about ancient stone.

PHOTO JOURNAL

P. 135 Hamelin Pool has another secret that Peter Kopke shared with us. The salty waters leach calcium carbonate out of millions of tiny beach shells. The calcium carbonate then combines with other minerals and silt to form thick layers of limestone.

Peter and his neighbors quarry the limestone to build their homes. It provides perfect insulation, keeping buildings cool in the brutally hot summers and warm in the mild winters.

Pp. 138–139 These stromatolites, covered in red algae, looked as if they had been transported from the Archeon Eon and planted here. They had a surreal shape and feel to them, so different from the other mats and columns that surrounded this group.

Sights like these gave us the feeling that there was a deeper history just out of our reach. Only later did we learn about the impressive legacy of the bacteria.

Pp. 142-143 We were intrigued by the jellyfish with their stubby tentacles and star-shaped designs revealed on their undersides when they turned over. The jellyfish seemed more like mascots of the bay than predators. Their only companions were tiny fish that grazed on the bacteria. Swift as silver darts, the fish were hard to photograph until we caught them in a shaft of light against the shadows.

We tried to get several pictures of the underwater stromatolites but couldn't focus the camera properly—the water was cold enough to numb our hands within minutes. We were lucky to get a few shots. In some winters, a thin sheet of ice can cover the shallow bay.

Pp. 144-147 Stromatolites like these reminded Lynn of the work of Jean Arp, a French-German sculptor whose rounded, organic forms are popular in Europe. Lynn tried to frame her shots to capture the same sense of play that Arp conveys in his sculptures. We joked that the bacteria may have gone through several "artistic periods" as they built the stromatolites, which would account for the variety of their formations.

ANCIENT MINERALS & ANCIENT LIFE WITHIN US

When we began our journey, we never thought it would lead us to the link between ancient stone, ancient life, and humanity. This understanding developed slowly, resonating deeper with each journey and each year as the book developed.

Pp. 150-151 We sent a sample of Acasta gneiss to a colleague whose grandfather cut and polished the stone and then photographed a close-up of its surface. The dense, layered mineral structure shows flecks of pink granite, feldspar, apatite, zircon, and other igneous minerals. The texture has the look and feel of a landscape swept by fire. Some of these ancient minerals, such as apatite, help form the structures within our bodies.

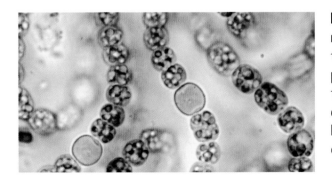

Pp. 158-159 Cyanobacteria, like this *Nostoc cyanobacteria*, not only helped to produce Earth's oxygen but also became the chloroplasts in plants, giving them the ability to use photosynthesis. Chloroplasts and mitochondria helped fuel the explosion of plant and animal life on Earth. Other types of bacteria perform essential functions within us that help keep us alive. How and when did bacteria first arise? That question remains the deepest mystery of all.

AFTERWORD

from L. Sue Baugh

Our encounters with these ancient sites were like stones dropped into the still waters of a deep pond. The ripples continue to well across our personal and creative lives, reawakening and informing ourselves and our artistic work. The two of us began this journey as visitors to the sites but ended by feeling we had come home.

For me, the ripples extend far beyond our personal journey. I believe humanity's intimate connection to Earth offers profound knowledge and imagery not just for artists but for everyone. Concepts of Deep Time, of Earth's story within our own bodies, of ancient places that recognize and welcome us—these underscore the mystery of our origins. Such knowledge, with its comfort and hope, is available for the listening.

The final blessing of our project came after our last trip, when we held a photo exhibit in Vevey, Switzerland. At the exhibit opening, we saw for the first time the transforming power of these images on other people. Yet we still carried lingering doubts about our authenticity as artists. Near the end of the evening, an older woman came up to us and asked if we were geologists.

"No," we said, "although we've always had a love of stones."

"Are you professional photographers then?"

"Not exactly. For the most part, we let the places decide where and when to take the pictures."

Her eyes lit up in sudden understanding; she leaned forward as if speaking directly to our doubts. "Ah! Then you are *artists!*"

> *You draw close to nature; your vision opens, and you begin to learn.*
> —David Mowaljarlai, Australian Aboriginal elder

ACKNOWLEDGMENTS

First, and most importantly, to my travel partner, *Lynn Martinelli*, who shared the joys and trials of the road and contributed her artistic insights and one third of the photos to the book. We are both deeply grateful to the following people who gave generously of their time, expertise, and support on our journeys to these remote and powerful sites.

Mt. Narryer, Western Australia: *Dr. David Nelson*, Curtin University of Technology, Perth, Western Australia; *Dr. John Valley*, Department of Geology, University of Wisconsin at Madison, Wisconsin; *Dr. Aaron Cavosie*, Department of Geology, University of Puerto Rico, Mayaguez, Puerto Rico; *Margaret Duff*, Electorate Officer, Government of Western Australia, Perth, Western Australia; *Sandy and Carol McTaggart*, Mt. Narryer Ranch, Mullewa, Western Australia, for their warm friendship and permission to stay at Mt. Narryer.

Hamelin Pool, Shark Bay, Western Australia: *Dr. Lindsay Collins*, Curtin University of Technology, Perth, Western Australia; *Dr. Stephen Mojzsis*, Professor of Geological Sciences, University of Colorado, Boulder, Colorado; *Peter Kopke*, Carbala Homestead at Hamelin Pool for his kind permission to camp on his beach.

Akilia Island, Greenland: *Dr. Stephen Mojzsis*, Professor of Geological Sciences, University of Colorado, Boulder, Colorado; *Dr. Minik Rosing*, chair of the Commission for Scientific Research in Greenland, director of the Geological Museum at the University of Copenhagen, Denmark; *Dr. Jurate Gertzbein*, Mineral Development Advisor, Iqaluit, Nunavut, Canada; *Dr. Ian Glasspool*, Department of Geology, Field Museum of Natural History, Chicago, Illinois; *Bjarne Kreutzmann*, former mayor of Nuuk, Greenland; *Ludvig Sethsen*, guide to Akilia Island, Tourism Office, Nuuk, Greenland.

Blacktail Canyon, Grand Canyon, United States: The *river guides* of Wilderness River Adventures, who not only saw us safely down the river but gave us a geology tour of Blacktail Canyon. Geologist *J. Greer Price*, whose book *An Introduction to Grand Canyon Geology*, provided a beautifully illustrated description of the rock layers in the canyon.

Acasta River and Great Slave Lake, Northwest Territories, Canada: District Geologists *Diane Baldwin and Karen Gochnaur*; *Walt Humphries*, minerologist; *Tim O'Loan*, Chief Aboriginal Negotiator; *Angus Charlo*, pilot for Air Tindi.

Special thanks to: *Judith Gallagher*, who kept saying these journeys had to be made and who meticulously read several versions of the manuscript; the Unfolding Group (*Julia, Carol, Diane, and Betty*) and the *Williamsburg Writers* for their unflagging encouragement; *Robert Hamper and Sue Roupp* for arranging speaking engagements; *Nancy Caffall*, retired geologist from University of Massachusetts at Amherst, who reviewed the science chapters; *Michael Francis*, for his design expertise; to my family and many friends who supported this work from the beginning.

To *Jericho Hernandez*, book designer, whose skill, patience, and good humor during the creation of this book are deeply appreciated. May every page be a portfolio page for you.

And especially to my husband, *Norm Zuefle*, whose patient critiques and emotional and financial support during the writing of this book helped make *Echoes of Earth* possible.

SUGGESTED READING

The following books offer a rich feast for the mind and senses on topics found in *Echoes of Earth*.

Origins: The Evolution of Continents, Oceans, and Life, by Robert Redfern. United Kingdom; Cassell & Co., 2000.

Prehistoric Life: The Definitive Visual History of Life on Earth, Angeles Bavira Guerrero and Peter Francis, eds. London: Dorling Kindersley, 2009.

The Raven's Gift: A Scientist, a Shaman, and Their Remarkable Journey Through the Siberian Wilderness, by Jon Turk. New York: St. Martin's Press, 2011.

Symbiotic Planet: A New Look at Evolution, by Lynn Margulis. Amherst: Massachusetts Sciencewriters, 1998. How bacteria have transformed Earth.

Within the Stone, by Bill Atkinson. San Francisco: Browntrout Publishers, 2004. Stunning close-ups of stones.

Yorro, Yorro, by David Mowaljarlai and Jutta Malnic. Broome, Western Australia: Magabala Book Aboriginal Cooperation, 1993. Mowaljarlai describes the Aboriginal worldview.

BIBLIOGRAPHY

Mt. Narryer, Western Australia

Braun, Jean, et. al. *Structure and Development of the Australian Continent: Geodynamic Series, vol. 26*. Washington, D.C.: American Geophysical Union, 1998.

Morrison, Reg, and Maggie Morrison. *Australia: The four-billion-year journey of a continent*. Sydney, Australia: Weldon Publishing, 1988.

Myers, John S. "Precambrian history of the West Australian craton and adjacent orogens." 1993. *Annual Review of Earth Planet Science* (21), 453–85.

Neld, Ted. *Ten Billion Years in the Life of Our Planet*. Cambridge, Massachusetts: Harvard University Press, 2007.

Nelson, D.R. "Geochronology of the Archean of Australia." 2008. *Australian Journal of Earth Sciences*,(55) 6, 779–793.

Young, G. C., and J. R. Laurie, eds. *An Australian Phanerozoic Timescale*. Melbourne: Oxford University Press, 1996.

Akilia Island, Greenland

Born, Erik W., and Jens Bocher. *The Ecology of Greenland*. Trans. Danny Eibye-Jacobsen. Ministry of Environment and Natural Resources: Nuuk, Greenland, 2001.

Greenland Atlas. Nuuk: Greenland: Ataukkiorfik Publishers, 1993

Henriksen, Niels. *Geological History of Greenland: Four billion years of Earth evolution*. Geological Survey of Denmark and Greenland (GEUS), Denmark, 2008. English Edition.

Nutman, Allen P., Stephen J. Mojzsis, and Clark R.L. Friend. "Recognition of 3850 Ma water-lain sediments in West Greenland and their significance for the early Archean Earth. "

1997. *Geochimica et Cosmochimica Acta* Vol. 61(12), 2474 – 2484.

Schopf, J. William. *Cradle of Life: The Discovery of Earth's Earliest Fossils*. Princeton, New Jersey: Princeton University Press,1999.

Blacktail Canyon, Grand Canyon, United States

Beus, Stanley S., and Michael Morales. *Grand Canyon Geology*. New York: Oxford University Press, Inc:, 1996.

Hamblin, W. Kenneth, and J. Keith Ribgy. *Guidebook to the Colorado River, Part 1: Lee's Ferry to Phantom Ranch in the Grand Canyon National Park. Studies for Students No. 4, Vol. 15, Part 5*. Brigham Young University Geology Series, 1968.

Hamblin, W. Kenneth, and J. Keith Ribgy. *Guidebook to the Colorado River, Part 2: Phantom Ranch in Grand Canyon National Park to Lake Mead, Arizona-Nevada. Studies for Students No. 5, Vol. 16, Part 2*. Brigham Young University Geology Series, 1969.

Price, L. Greer. *An Introduction to Grand Canyon Geology*. Grand Canyon, Arizona: Grand Canyon Association, 1999.

Stevens, Larry. *The Colorado River in Grand Canyon: A comprehensive guide to its natural and human history*. 6th ed. Flagstaff, Arizona: Red Lakes Books, 1983.

Whitney, Stephen R. *A Field Guide to the Grand Canyon*. 2nd ed. Seattle, Washington: The Mountaineers, 1996.

Acasta River, Northwest Territories, Canada

Bastedo, Jamie. *Shield Country: The life and times of the oldest piece of the planet*. Alberta, Canada: Red Deer Press, 1994.

Bowring, Samuel A., and Ian S. Williams. "Priscoan (4.00-4.03 Ga) orthogneisses from Northwestern Canada." 1999. *Contributions to Mineralogy and Petrology*. Vol. 134, 3–16.

Science 321(5897), 1828. [DOI: 10.1126/science.1161925].

Great Slave Lake, Northwest Territories, Canada

Knoll, Andrew H. *Life on a Young Planet: The first three billion years of evolution on Earth*. Princeton, New Jersey: Princeton University Press, 2003.

Konhauser, Kurt. *Introduction to Geomicrobiology*. Oxford, UK: Blackwell Science Ltd:, 2007.

Lane, Nick. *Oxygen: The Molecule that Made the World*. New York: Oxford University Press, 2002.

Margulis, Lynn, and Dorion Sagan. *Microcosmos: Four billion years of microbial evolution*. Berkeley and Los Angeles, California: University of California Press, 1997.

Margulis, Lynn, and Dorion Sagan. *What Is Life?* New York: Simon & Schuster, 1995.

Microbes@NASA. National Aeronautics and Space Administration. http://microbes.arc.nasa.gov/about/stomatolites.html

Schopf, William J. *Cradle of Life: The discovery of Earth's earliest fossils*. Princeton, New Jersey: Princeton University Press, 1999.

Shark Bay, Western Australia

Konhauser, Kurt. *Introduction to Geomicrobiology*. Oxford, UK: Blackwell Science Ltd:, 2007.

Morrison, Reg, and Maggie Morrison. *Australia: The four-billion-year journey of a continent*. Sydney, Australia: Weldon Publishing, 1988.

Southwood, Richard. *The Story of Life*. New York: Oxford University Press, 2003.

Wilson, Edward O. *The Diversity of Life*. Rev. ed. New York: W.W. Norton & Company, 1999.

Ancient Minerals Within Us

Anderson, H. Clarke, Rama Garimella, and Sarah E. Tague. "The role of matrix vesicles in growth plate development and biomineralization." (January 1, 2005), *Frontiers in Bioscience*, vol. 10, 822–837.

Angier, Natalie. "Bone, a Masterpiece of Elastic Strength." *The New York Times*. April 28, 2009. www.nytimes.com/2009/04/28/science.

"Apatite." www.answers.com/topic/apatite

Seeley, Rod R., Trent D. Stephens, and Philip Tate. "Chapter 06: Skeletal System: Bones and Bone Tissue," *Anatomy and Physiology*, 8th ed. New York: McGraw-Hill Companies, 2008.

"Structure of bone." DoITPoMS Teaching and Learning Packages. University of Cambridge, 2009. http://www.msm.cam.ac.uk/dolpoms/tiplib/bones/structure.php

Uthman, Ed., MD. Diplomat, American Board of Pathology. "Elemental composition of the human body." Feb 14, 2000 http://web2.airmail.net/uthman/elements_of_body.html

Ancient Life Within Us

Cavalier-Smith, Thomas. "Origin of mitochondria by intracellular enslavement of a photosynthetic purple bacterium." 2006. *Proceedings of the Royal Society of London B*. vol. 273 (1596), 1943-1952.

Davidson, Michael W. "Mitochondria." Molecular Expressions™-- Cell Biology and Microscopy, Structure and Function of Cells & Viruses. 1995-2009, Florida State University. http://micro.magnet.fsu.edu/cells/mitochondria/mitochondria.html.

Lang, Franz, et. al. "On the origin of mitochondria and Rickettsia-related eukaryotic endosymbionts." 2005. *Jpn. J. Protozool.*, vol. 38(2), 171-183.

Margulis, Lynn. *Symbiotic Planet: A New Look at Evolution*. Amherst, Massachusetts: Sciencewriters, 1998.

Margulis, Lynn, and Dorion Sagan. *What Is Life?* New York: Simon & Schuster, 1995.

Martin, William. "A briefly argued case that mitochondria and plastids are descendants of endosymbiots, but that the nuclear compartment is not." 1999. *Proceedings of the Royal Society of London B*. vol. 266(1426), 1387-1395.

Martin, William, et. al., "Evolutionary analysis of Arabidopsis, cyanobacterial, and chloroplast genomes reveals plastid phylogeny and thousands of cyanobacterial genes in the nucleus." www.pnas.org/conent/99/19/12246.full.

Strachan, Tom, and Andrew P. Read. "Our place in the tree of life." *Human Molecular Genetics*, 3rd ed. Oxford, UK: Garland Science/Taylor & Francis Group, 2003.

Art & Science of Earth History

Manning, Craig E, Stephen J. Mojzsis, and Mark T. Harrison. "Geology, Age and Origin of Supracrustal Rocks at Akilia, West Greenland." 2006. *American Journal of Science*, vol. 306; 303-366.

Margulis, Lynn and Dorion Sagan. *What Is Life?* New York: Simon & Schuster, 1995.

McNamara, Ken. *Stromatolites*. Perth, Australia: Western Australia Museum, 1992.

Morgan, Kendall. "A rocky start: Fresh take on life's oldest story." 2003. *Science News* (163), 264–266.

O'Neil, Jonathan, Richard W. Carlson, Don Francis, and Ross K. Stevenson. "Neodymium-142 evidence for Hadean mafic crust." (26 September 2008) *Science News*.

"Radiometric Dating." Evolution 101: Molecular Clocks. http://evolution.berkeley.edu/evosite/evo101/

Witze, Alexandra. "Grand Canyon's high surroundings may be product of continental lift." (May 21, 2011) *Science News* (179), 12.

Valley, John W. , William H. Peck, and Elizabeth M. King. "A cool early Earth." 2001. *Geology*, vol. 30(4) 351–354.

Timelines of Earth

"The Archeon Eon I: The Oldest Rocks, and Eon II: The Origin of Life." Geol 102 Historical Geology. Spring 2001. http://www.geol.umd.edu/~tholtz/G102/102Lec17.htm

Brandt, Niel. "Geological/evolutionary time." http://eathsci.org/fossils/geotime/time/geotime2.htm

Colebrook, Michael. "Chronology of Earth History." http://www.greenspirit.org.uk/resources/chronology.html

"Geologic Time Scale." www.geo.ucalgary.ca/~macrae/timescale/timescale.html

Gore, Pamela J. W. "Introduction to the Early Paleozoic." Georgia Perimeter College. http://www.dc.peachnet.edu/~pgore/geology/geo102/cambrian.html

Guerrero, Angeles Gavira, and Peter Frances, eds. *Prehistoric Life: The Definitive Visual History of Life on Earth*. London: Dorling Kindersley, 2009.

"Lecture 7: Earth's First 3.7 Billion Years." http://rainbow.ldgo. columbia.edu/courses/v1001/7.html

Palmer, Douglas, and Peter Barrett, illustrated. *Evolution: The Story of Life*, Berkeley: University of California Press, 2009. First printed in England by Mitchell Beazley, an imprint of Octopus Publishing Group, 2009.

Rollinson, Hugh. "The Origin of Life." Funded by the National Subject Centre for Geography, Earth & Environmental Science. August 2001.

Siry, Joseph. "The Geological Record." http://web.rollins. edu/-jsiry/CREATION.html

Schlesinger, W. H. *Biogeochemistry: An Analysis of Global Change*. Chicago: Academic Press, 1996.

Taylor, Thomas N., Edith L. Taylor, and Michael Krings. *Paleobotany: The Biology and Evolution of Fossil Plants*. Boston: Elsevier, 2009.

Wacey, David. *Early Life on Earth: A Practical Guide*. New York: Springer, 2009.

PHOTO CREDITS

Mt. Narryer, Western Australia pp. 2-3 Western Australia, Jacques Descloitres, MODIS Land Rapid Response Team. NASA/GSFC; **pp. 5-7** Aerial view of Mt. Narryer © Reg Morrison, 2001; **p. 8** © Rand McNally; **p. 9** Zircons in polarized light © Reg Morrison; **pp. 28-29** Southern Sky, © Christopher J Picking, www.starrynightphotos.com; **p. 30** Photo courtesy of Sandy and Carol McTaggart, 2011.

Akilia Island, Greenland pp. 32-33 Greenland, NASA World Wind software; **p. 38** © Rand McNally.

Blacktail Canyon, Grand Canyon, United States pp. 60-61 Grand Canyon, United States, MODIS Land Rapid Response Team, NASA/GSFC; **p. 66** © Rand McNally

Acasta River, Northwest Territories, Canada pp. 80-81 Northwest Territories, Canada, Jacques Descloitres, MODIS Land Rapid Response Team. NASA/GSFC; **p. 86** © Rand McNally; **p. 105** Lichen close-up © David Aubrey/Photo Researchers, Inc.

Great Slave Lake, Northwest Territories, Canada pp. 110-111 Great Slave Lake, Northwest Territories, Canada, Jacques Descloitres, MODIS Land Rapid Response Team. NASA/GSFC; **p. 116** © Rand McNally; **p. 120** Fossil cyanobacteria, source unknown; Oxygen bubbles from cyanobacteria © Reg Morrison; Stromatolites on a Young Earth © L. Sue Baugh 2011.

Shark Bay, Western Australia pp. 128-129 Shark Bay, Western Australia, Liam Gumley, University of Wisconsin – CIMM; **p. 134** © Rand McNally.

Ancient Minerals Within Us pp. 150-151 Acasta gneiss close-up, courtesy of Ernie Lombard; **p.152** Hydroxy-apatite crystal © The Internet Encyclopedia of Science; **pp. 153-155** Haversian canals in compact bone © Andrew Syred/Photo Researchers, Inc.; **p 156** Illustration of hip bones, © L. Sue Baugh 2011; **p. 157** Illustration of human pelvis and hip bones © Sioban Lombardi 2010; Cross section of human hip bone, Macrophotograph,

magnification: X 3 (6x7), X 1.3 (35mm) © Freidrich Michler/ Photo Researchers, Inc ; Compact bone lamellae, False-color SEM magnification: x1200 at 6x7cm size © Professor Pietro M. Motta/Photo Researchers, Inc ; Spongy bone. Colored SEM © Susumu Nishinaga/Photo Researchers, Inc; Illustration of apatite needles, © L. Sue Baugh 2011.

Ancient Life Within Us pp. 158-159 Nostoc cyanobacteria © Dr. Robert Calentine/Visuals Unlimited/Getty Images; **p. 160** Mitochondria in human cell, magnification: x6380 at 6x7cm size. © Professors P. M. Motta, S. Makabe & T. Naguro/Photo Researchers, Inc.; **pp. 160** Heart illustration, © L. Sue Baugh 2009; **pp. 161-163** Cardiac muscle tissue with mitochondria, Colored SEM © Steve Geschmeissner/Photo Researchers, Inc. **p. 165** Human body, © Sioban Lombardi 2010; Bacteroides © Dennis Kunkel Microscopy/ Visuals Unlimited, Inc.; Bacteroides fragilis, false color TEM © CNRI/Photo Researchers, Inc.; Escherischa coli, false-color SEM mag. X2,350 at 6x7 cm size © P. M. Motta & F. Carpino/Univ. "La Sapienza"/Photo Researchers, Inc.; Lactobacillus salivarius bacteria, © Scimat/Photo Researchers, Inc.

Art & Science of Earth History "Earth's Oldest Minerals" **p. 168** Zircon crystals in polarized light © Reg Morrison; **p. 169** (bottom left) Zircons relative size and (top right); Zircon in x-ray crystallography © Aaron Cavosier, Univ. of Puerto Rico; **p. 170** "Primitive Earth" © David Hardy, www. astroart.com; "Cool Early Earth," © Don Dixon, www.cosmographica.com/gallery.

"Earth's Oldest Skin" **pp 171, 172-173** Illustrations, © L. Sue Baugh, source material, see Bibliography p. 199

"Earth's Oldest Colonies of Life" **p. 174** Building of stromatolite © L. Sue Baugh 2011; **p. 175** Photo of fossil stromatolite, Western Australia © Reg Morrison.

Typical Timeline of Earth pp. 178, 183 Illustrations © L. Sue Baugh 2011, source material, see Bibliography, pp. 199-200

Full Timeline of Earth p. 179 Spiral Galaxy NGC 4414, Hubble Heritage Team; Oldest known zircon, 4.4 byo, from Jack Hills, W. Australia, photo courtesy Dr. J.W. Valley, University of Wisconsin at Madison; Polarized zircons from Mt. Narryer © Reg Morrison; Illustration of solar system © Sioban Lombardi 2011; Computer illustration of Human DNA © hybrid medical animation/Photo Researchers, Inc.; Moon photo © NASA; "Cool Early Earth" © Don Dixon; **p. 180** Rock-eating slime bacteria © T. Stevens & P. McKinley, PNNL/Photo Researchers, Inc.; Prokaryote illustration © Sioban Lombardi 2011; Isua Greenstone Specimen Li9223, photograph courtesy of Dr. Ian Glasspool, the Field Museum of Natural History, Chicago, Illinois; Fossil stromatolite © Reg Morrison; Gunflint chert, Ontario, Canada © Sinclair Stammers/Photo Researchers, Inc.; Oxygen bubbles from cyanobacteria © Reg Morrison; Banded iron formation © Fossil Mall, Inc., www.fossilmall.com; **p. 181** Eukaryote illustration © Sioban Lombardi 2011; Marine red algae © Biophoto Associates/Photo Researchers, Inc.; Ciliate protozoa mag x180 at 10 cm wide © Laguna Design/Photo Researchers, Inc.; **p. 182** Paramecium caudatum reproducing © Michael Abbey/ Photo Researchers, Inc.; Precambrian life illustration © Sioban Lombardi 2011.

INDEX

ABOUT THE ARTISTS

 *L. Sue Baugh* grew up surrounded by the woodlands and open fields of rural Illinois. From these landscapes, she found her early themes as a writer and painter. In college, she focused on creative writing, earning a BA from the University of Iowa and an MFA from the University of North Carolina. Since then, she has worked in the publishing industry as a writer/editor in language arts, history, and science.

In early 2001, she conceived of the ancient rocks project. "I wanted to listen to the land again, to recapture something that over the years I had lost from my life." The experiences and knowledge she gained on the journey re-ignited her artistic life, enabling her to create the unique features of *Echoes of Earth*. "I wanted readers to have their own direct experience of these sites. I know how much my own work has changed because of our encounters with the stones. They have a lot to say!" She currently lives in the Chicago area with her husband.

Lynn Martinelli spent time exploring the outdoors as a child and imagining faraway places. Those dreams eventually became reality as she began traveling, studying, and working abroad, finally settling in Switzerland. In college, she concentrated on art, writing, and languages—the basis of her career in communications and technology. "In my early paintings—inspired by ancient maps, cave paintings and tapestries—I layered printed pixels and hand markings on paper scrolls that could not be contained in a frame."

During a trek in the virgin rainforest of Borneo, her childhood love of nature blossomed into a sense of oneness with all life on the planet. This connection deepened in surprising ways through the ancient rocks project. "Opening creatively to these ancient rock sites resulted in a profound artistic and personal transformation. The stones continue to be a well of inspiration, on many levels. When they talk, I listen!" She currently lives in Basel, Switzerland.

ABOUT THE PRINTER

We chose Hung Hing Printing Group in China to produce *Echoes of Earth* because of their reputation for their eco-friendly practices, their high-quality printing, and their ability to produce all our special layouts. Hung Hing has earned certifications in employee safety and well being, in using sustainable resources, and in ensuring responsible manufacturing practices. They continually seek ways to improve their business practices.